# BIRDS IN TROUBLE

# BIRDS IN

TEXAS A&M UNIVERSITY PRESS • *College Station*

# TROUBLE

LYNN E. BARBER

This paper meets the requirements
of ANSI/NISO Z39.48-1992 (Permanence of Paper).
Binding materials have been chosen for durability.
Manufactured in China by Everbest Printing Co.
through FCI Print Group

LIBRARY OF CONGRESS CATALOGING-IN-PUBLICATION DATA

Barber, Lynn E., author.
    Birds in trouble / Lynn E. Barber. — First edition.
        pages cm
    Includes bibliographical references and index.
    ISBN 978-1-62349-359-2 (flex with flaps: alk. paper) —
    ISBN 978-1-62349-360-8 (ebook)
1. Rare birds—United States.  2. Birds—Conservation—United States.  I. Title.
    QL676.7.B37  2016
    598.168—dc23
                        2015018637

For my husband, David,

who gave me the idea for this book

and has encouraged my passion

for birds. Thanks also to Shannon Davies,

Texas A&M University Press, without whom

this book would not have happened.

# C O N T E N T S

# P R E F A C E

When this book was just a dream beginning to take shape in my mind, oil was washing up on the shores of Louisiana, covering Brown Pelicans, their nests, and their eggs. Whooping Cranes had recently suffered a dramatic die-off due to lack of food at their wintering spots in coastal Texas. It was spring, and migrating shorebirds were once again facing a reduction in their breeding habitats as they returned to the United States from the south.

So many bird species (as well as other animals and many plants) are declining in numbers, year by year. Many bird species are at, or are nearing, such low numbers that their ultimate fate is uncertain at best. Serious, thoughtful books have been written about this problem, many people are working to make a difference, and ways to stop or reverse the trends in the different species have been suggested. But many of the problems persist and worsen.

Although environmental problems, like many other major, long-standing problems around the world, are not likely to be solved by any one of us or by any group of us or to be solved quickly, it is worthwhile to try to work toward solutions. There are things we can do that can make a difference. This book is aimed at encouraging each of us to do something, to take action, to be part of improving the plight of birds, both those that are endangered and those that are around our homes and parks every day. I hope that each of you will try to learn about birds, about the problems, and about what can be done, so that we all will be on the side of the environment, rather than giving up without trying or, even worse, unashamedly being part of the problem.

There are four main sections in this book. The first section, "The Importance of Bird Habitat," provides general information on some of the main habitats that are important for birds, as well as examples of how birds have been and are being affected by changes in these habitats. The next section, "What Can Be Done to Help Birds," gives examples of general actions that people can take to help birds, including actions that can benefit endangered birds. The "Species Accounts of Birds in Trouble" section aims to make you, the reader, aware of selected examples of birds in the United States that are in trouble and the likely reasons for the trouble, and it presents ways for all of us, as individuals and as groups, to figure out what we might do (or refrain from doing) to improve the situation. Most of the birds included in this section of the book are listed by one or more federal or state agencies or organizations as being "endangered," "threatened" or otherwise in dire straits, or headed in that direction. The final section of this book provides information and ideas on how each of us can help the everyday birds in our yards and in our communities, particularly in view of the habitat stresses and problems that birds face.

Birds that are extinct (Passenger Pigeon) or very likely extinct (Ivory-billed Woodpecker) or hanging on by a thread (California Condor) are not discussed in detail here, nor are most bird species that only infrequently appear in the United States (such as wandering Asian or Mexican species) or have only a small portion of their worldwide population in the United States (such as most of the parrots and parakeets that have small established US populations derived from escaped cage birds).

For readers who may not have as much prior knowledge of birds as amateur experts with long life lists, each bird account in this book presents a brief background of what the species looks like, how it behaves, and what its habitat preferences are. The historical abundance and current status of each of the selected species are discussed, as well as the reasons to be concerned about each species, the likely causes for low numbers and/or decline in their numbers, and the problems that each species faces or appears to face.

I hope that the accounts of birds in trouble will serve as a beginning for you and that you will move beyond these brief accounts to learn even more about the birds discussed here and their environments, as well as about the hundreds of other bird species that are of concern. I also hope that your caring about birds in trouble will help you try to make, and actually make, a difference in their, and our, world.

## THE DREAD OF EXTINCTION

Among people who pay attention to birds are many birders whose love of the chase, as well as the amount of time spent on it and their bird tally list, leaves them little time or energy for conservation. In contrast are the bird lovers for whom conservation is an integral and all-consuming part of making sure that there are birds to watch now and in the future. In between these extremes are those who try to be both active birders and active conservationists, who perhaps would rather be listing and chasing birds but who have come to the realization that unless something is done now and in the future to improve bird habitats and change our society's and our own individual habits, we will be caught in the downward spiral of a "Silent Spring" from which there will be no return for many species. The Passenger Pigeon is symbolic of what we dread: birds that once were so numerous they blackened the skies, and then were gone before anyone was able to take measures to keep them from extinction.

## WHY CARE?

Most people who believe that it is critical to keep bird species from extinction wrestle with how to persuade others of the importance of this mission. How can we place a monetary value on a species that our progress- and bottom-line-oriented commercial society will recognize? For those who are part of a religious tradition that includes a Creator or values life and living beings, animals and plants are to be treasured and used wisely, without destroying or harming the creation if at all possible. We must be concerned about the "least of these" in the creation.

Even apart from religion, those who are knowledgeable about the environment and its plant and animal components know that destruction or harm of any part of the environment affects the whole, often in a way that is at best unpredictable and at worst devastating. While we humans often feel removed and apart from that whole, we are not. Casting our pollution into the waters or the air or onto the land clearly can cause severe problems for us, affecting our health, our crops, our recreational activities, and the world we smell and taste and hear and see when we look around us.

Part of the problem of course is that, if we have changed the environment so much that a single species is endangered, there is a sort of "slippery slope" effect. Once we have endangered or lost that species, later events occur that cause further changes that cannot be stopped or are extremely expensive or difficult to stop or reverse. This is often thought of as a "butterfly effect," which is a term coined by Edward Lorenz to indicate a sensitive dependence on initial conditions, where one tiny change to an initial condition (e.g., a butterfly's wings moving) can later result in a huge change (e.g., a hurricane).

As caretakers of the world, we should take an oath analogous to the physician's Hippocratic Oath: "First, do no harm." It is a great challenge to all of us who are concerned about the environment to help others see that allowing a species to slide or plunge into extinction not only does harm to the species and its environment but also does harm to our world as a whole, as well as to our bodies and our souls.

### HOW MANY BIRDS ARE LEFT?

It is very difficult, for many reasons, to know exactly how many bird species there are now, and it is even more difficult to know how many there were a few years ago and, even more so, when early naturalists lived. The numbers presented herein at the beginning of the discussions of the individual species and in appendix 1 are estimates of how many birds there are of each species. These estimates come from the National Audubon Society's 2007 Watchlist, as well as from various references that discuss the particular species. Where there seemed to be significant differences between estimates, I have included a range of numbers.

### WHAT IS AN "IN-TROUBLE" BIRD?

In this book, I have generally used the phrase "in trouble," rather than such terms as "endangered," "threatened," "vulnerable," and the like, when discussing a species. Various organizations and governmental agencies use those and other terms and give them particular definitions that do not always agree or may overlap.

Appendix 2 provides the criteria for classifying birds as "in trouble" according to various agencies: the International Union for the Conservation of Nature and Natural Resources (IUCN), the American Bird Conservancy, and the National Audubon Society (NAS). The NAS periodically issues its Watchlist, as do various state chapters of the National Audubon Society. As shown in appendix 2, the NAS Watchlist categorizes species that are in trouble as "red" (declining rapidly and/or having very small populations or limited ranges and facing major conservation threats; such species are typically of global conservation concern) and "yellow" (species that are either declining or rare; these species are typically of national conservation concern).

For the birds in trouble described in this book, appendix 3 provides the current status under the Endangered Species Act. Comparing appendixes 2 and 3 makes it clear that there are many species "in trouble" that are not yet protected under the Endangered Species Act.

Although federal and state definitions of the status of birds may vary, it is useful to know those of the US Fish and Wildlife Service:

E = Endangered: in danger of extinction throughout all or a significant portion of its range

T = Threatened: likely to become endangered within the foreseeable future throughout all or a significant portion of its range

C = Candidate: under consideration for official listing when there is sufficient information to support listing

SC = Species of concern: no petitions for listing have been filed; have not been given E, T, or C status but have been identified as important to monitor

For more information see http://www.fws.gov/midwest/endangered/glossary/index.html.

Whether a particular species is placed in one or more of these categories by an organization or a governmental entity varies over time and from place to place across the United States and Canada. Therefore, I use the term "in trouble" to indicate that a species is not doing well, whatever its official or unofficial categorization. It may be a species that has always been too low in numbers to be safe from the possibility of extinction, a species that is decreasing or has decreased precipitously or gradually, or a species that is having difficulty reaching a sustainable level.

### BIRDS SELECTED FOR INCLUSION IN THIS BOOK

The birds discussed in this book as being "in trouble" are either included in *Birder's Conservation Handbook: 100 North American Birds at Risk*, have a global population of less than twenty-five thousand, have a substantially decreasing continental US population, have recently been or are now close to extinction, and/or are listed by one or more governmental agencies or conservation organizations as being endangered, threatened, or otherwise in trouble. While there are birds that fit into one or more of these categories and are not included in this book, an increased understanding of the issues that are raised in the accounts presented here will show the reader how to take actions that can be of benefit to other birds that are also in trouble.

In any case, each species described in this book has been selected because of concerns that its status must be carefully watched to be sure that it does not begin to decline or increase its rate of decline due to the various pressures that threaten it. For most of the species selected for this book, threats are real and ever present, and there is a need to DO SOMETHING NOW. What that something is varies from species to species. In some cases, the threat seems almost impossible to handle and the situation is dire, while in others there are concrete, realistic things that can and should be done to improve the situation incrementally or dramatically.

In a couple of cases, I have included species that currently have a relatively high population but are still deemed "in trouble" for some reason. For the Snowy Plover, for example, there are discrete populations in the United States that are on the brink of extinction, and for the Cerulean Warbler, the necessary forest nesting habitat is rapidly decreasing.

Some of the birds that are in trouble but not included in this book are certain subspecies or races that may be in trouble only in a particular geographic area where they were historically found in much greater numbers (e.g., the western Bell's Vireo is an endangered subspecies of a species that as a whole is not in danger). Though the Red Knot species as a whole is not in danger, it is included in this book because the particular population that migrates through the eastern United States has decreased dramatically. Also, some species that are in trouble in North America but are doing well elsewhere, such as in Latin America or Europe, are not included here, primarily to keep this book of a more reader-friendly size. To learn more about these other in-trouble birds, you might peruse some of the works in the Suggested Reading section of this book or search the Internet, particularly at sites mentioned in the main text and in the appendixes.

While not all of the birds covered in this book are listed as federally endangered, there are concerns, sometimes official concerns, about each of these species' future. Of course, whether or not a particular species is considered endangered or threatened, most of these species are still protected under the law during all or most of each year.

Upon reading the species accounts presented here, you may notice that many of those chosen for inclusion are not the "cute," colorful birds that tend to attract or be noticed by most people, and many of them are found in areas that are remote from most of us. While it was tempting to include more of the attractive, familiar birds, most of these species are not in danger. It is still very important that the habitat these common birds need for breeding, wintering, and migrating not be destroyed. Admittedly, however, while the Ivory

Gull was included primarily because the portion of the world population that is found in the Arctic region of the United States is in trouble, it is likely that its attractiveness played a part in the selection process. Some of the birds included in this book, such as the shorebirds, are plain, being brown and white or black and white, and thus are not particularly distinctive or noteworthy in appearance to a nonbirder. I have included them in the hope that they and their plight will become known to more people and that these birds will be valued and conserved in spite of their seeming drabness.

Readers should also note that birds found only or primarily in Hawaii are not included in this book. Of course, many of these Hawaiian birds are endangered or threatened, and some are close to or at extinction. These birds also urgently need to be protected. We can support organizations and efforts dedicated to protecting and conserving Hawaiian birds that are in trouble and the habitat where these birds are found. Examples include the American Bird Conservancy (http://www.abcbirds.org/abcprograms/oceansandislands /hawaii.html) and the Hawaii Audubon Society (http://www.hawaiiaudubon .org/). More information about these birds and their status can also be found in bird field guides devoted to Hawaiian birds.

### LAWS, TREATIES, RULINGS, AND AGREEMENTS RELATED TO BIRDS IN TROUBLE

While there is an almost unlimited number of enactments and documents related to birds, the environment, and protection of endangered birds, there are a few that deserve special mention.

The Migratory Bird Treaty Act (MBTA) between the United States and Canada went into effect in July 1918. This treaty makes it unlawful to pursue, hunt, take, capture, kill, or sell any of more than eight hundred species of migratory birds in the United States and Canada. In addition, the MBTA protects dead birds and bird parts, such as feathers, as well as eggs and nests. There are similar treaties between the United States and Mexico, Japan, and Russia. A "migratory bird" under the terms of the MBTA is any species or family of birds that lives, reproduces, or migrates within or across international borders at some point during its annual life cycle and includes many species that are currently legally hunted as game birds. Most birds are deemed "migratory" under this act. There can be criminal penalties under this act even for the accidental killing of a migratory bird.

The Endangered Species Act (ESA) has been in place since 1973. It is administered by the US Fish and Wildlife Service (FWS) and the National Marine Fisheries Service, which is overseen by the Commerce Department. Its main provisions are designed to conserve threatened and endangered plants and animals and the habitats where they are found. Animals and plants that are listed under the ESA are either "endangered" (in danger of extinction throughout all or a significant portion of their range) or "threatened" (likely to become endangered within the foreseeable future). In order to be listed under the ESA, animals or plants must be nominated by one of the enforcing agencies or by petition by anyone. If the particular agency deems that the listing is warranted, it undertakes a status review, after which it must decide within one year whether the species is to be listed or rejected for listing or if the agency is to be given more time to decide. After time allowed for proposal of a rule and a decision whether or not to actually list the species, the agency must also designate the critical habitat for each species. Listing decisions are subject to judicial review and can be overturned. The official FWS list for the United States is at http://www.fws.gov/endangered/species/us-species.html. A worldwide list of endangered species is also kept by the FWS.

The ESA makes it illegal to "take" a listed animal, and a "take" includes significantly modifying the habitat of that animal. The ESA also requires federal agencies to ensure that their actions are not likely to jeopardize the continued existence of a listed species or result in the destruction or modification of a critical habitat. The ESA is to be implemented by the Fish and Wildlife Service and the National Oceanic and Atmospheric Administration (NOAA) Fisheries Service.

Because the ESA has great potential to affect private property, there has been much debate about it, as well as a number of lawsuits. Two of the most famous legal cases involve the snail darter (a fish whose habitat is affected by the Tellico Dam on the Tennessee River) and the Spotted Owl (listed as endangered in 1990, resulting in millions of acres of Pacific Northwest forest becoming protected areas).

More recently, in September 2011, a federal judge in Tucson, Arizona, seeking to speed ESA protection, approved a legal agreement between the Center for Biological Diversity and the Fish and Wildlife Service that requires the FWS to make initial or final decisions on adding 757 different organisms (animals and plants) to the endangered species list by 2018.

Other laws of importance to the conservation of birds include international treaties such as the Convention on International Trade in Endangered Species of Wildlife Fauna and Flora (CITES), which was signed by 118 countries, including the United States (http://www.cites.org/). Species listed under CITES appendix I cannot be bought or sold for profit, while those listed under CITES appendix II can be bought or sold only if such trade does not harm the survival of the species. The Wild Bird Conservation Act (WBCA) of 1992 (https://www.fws.gov/international/laws-treaties-agreements/us-conservation-laws/wild-bird-conservation-act.html) relates to birds mentioned in the appendixes of CITES and lists the requirements that must be met to obtain a permit to import a bird that legally may be imported.

# BIRDS IN TROUBLE

# THE IMPORTANCE OF BIRD HABITAT

This section includes information on some of the main types of habitats that are important to birds in general, as well as the habitat requirements of some of the endangered birds discussed in this book.

## WETLANDS

Wetland habitats are important for a wide variety of bird species. Waterfowl, such as ducks, geese, and swans, need lakes and ponds for breeding and for feeding. A serious problem for waterfowl such as Trumpeter Swans is the draining of rivers, lakes, and marshes so that the land can be used for farming and ranching, which thus makes the area unsuitable for waterfowl. Other wetlands have been polluted. Many lakes and rivers are used for recreation and have become surrounded by subdivisions.

In order to increase success rates for waterfowl, it is important to set aside substantial acreage consisting of water-filled breeding habitat, such as the remote western and northern wetlands in the United States and Canada. Trumpeter Swans and other waterfowl also need this type of habitat in which to winter. Agricultural activities are often of benefit to the swans in winter or when they are on migration because they provide places for the swans

Trumpeter Swan

to forage, especially if there are wetlands in the area. Success is even more likely if the wetlands are connected to and not remote from other wetlands. In some cases, preserving such links may mean undoing previous wetland destruction. In addition, the more wetlands from which lead shot is removed, the better, and enforcing hunting regulations where they exist, including the banning of lead shot, is also important. On a more global scale, reducing global warming, which is responsible for the rise in sea levels that reduces coastal swan nesting habitat, is important for the swans, as well as for other wildlife.

In addition to ducks, geese, and swans, birds that breed in wet meadows and other areas that have fresh or brackish water are negatively affected by loss of suitable wetland habitat. Human activity in and around wetlands and loss of wetland habitat are the biggest problems facing Yellow Rails. In many areas where Yellow Rails either breed or winter, wetlands are being destroyed by commercial and residential development, encroaching shrubs and trees, and drainage or channelization projects to transform wetlands into lakes or to make land suitable for crops. In habitat management decisions, there is often the need to balance competing habitat requirements of different species. Thus, increasing lake size for waterfowl can be detrimental to Yellow Rails, which require wetlands with more shallow water and more cover.

Wetlands that are used by Yellow Rails need to be managed to increase emergent vegetation where the rails nest. In other areas, prescribed burns and mowing are useful for reducing the growth of shrubs and trees in wetland areas. Because it is so difficult to know where Yellow Rails are, more research is needed to help determine their exact status and to learn more about what can be done to improve their habitat. It is particularly important that birders, desperate for a "lifer" view of the hard-to-find Yellow Rail, not trample wetland vegetation or destroy nests in breeding areas.

Wetland habitat is also managed for Whooping Cranes at their wintering site in Texas and along their migratory pathway. In Texas, freshwater ponds have been

Yellow Rail

constructed at Aransas National Wildlife Refuge to provide drinking water when the brackish waters become too salty. Prescribed burns are conducted to make the habitat better for cranes. Along the Platte River in Nebraska the grasslands are controlled by burning or cutting to reduce the height of the vegetation and to increase insect populations, both of which make the areas better for the cranes. There are plantings of agricultural crops specifically for the migrating cranes, both the Whooping Cranes and the far more numerous Sandhill Cranes.

## BEACHES

Beach habitat is important for many migrating birds, as well as some shorebirds that breed there. Snowy Plover populations near the tidal waters of the Pacific Ocean are particularly in trouble due to degradation of the beaches, oil spills that coat the birds' feet and plumage, beach recreation, beachfront development, and invasive plant growth, as well as increases in nest predators such as rats and gulls that are more tolerant of human activities than are the Snowy Plovers. Because the Snowy Plover is small and sand colored so that even well-meaning people can inadvertently approach the birds and their nests, uncontrolled beach vehicle and foot traffic can cause severe problems where the plovers nest. Nests are often abandoned if human or dog activity is too frequent or substantial. Where beaches are regularly raked to remove trash or kelp, they become less suitable for nesting and may have reduced food sources for the Snowy Plovers.

Pollution from oil spills in the Gulf of Mexico off the coasts of Florida, Louisiana, and Texas has exacerbated habitat problems already present there. In California and Oregon, invasive red foxes escaped or released from fur farms have been a major factor in the destruction of Snowy Plover nests. The interior Snowy Plover populations in Oklahoma, Texas, and Kansas have also had a reduction in breeding habitat due to flooding of salt flats and growth of invasive salt cedar plants, which also can hide predators that eat plover eggs and chicks.

While the Migratory Bird Treaty Act helped the numbers of Piping Plover recover in the early twentieth century, ongoing beach development and recreational activities have reduced usable habitat for this species. The Piping Plovers that breed in the northern United States and Canada migrate to eastern coastal beaches, and development, artificial adjustment of water levels

Snowy Plovers

on inland lakes and rivers, and other human activities have caused the birds to abandon their nests or to be displaced, thus reducing their nesting and wintering range. The Great Lakes, Northern Great Plains, and Atlantic populations of the Piping Plover have all been listed as endangered or threatened, with breeding birds having disappeared from Illinois, Indiana, Ohio, Pennsylvania, and portions of Ontario and Quebec. Although there have been increases in their numbers along the Great Lakes, these birds represent less than 2 percent of the species population. As with most shorebirds, protection of Piping Plover habitat is most important. Thus, regulating access to nesting areas, as well as beach development and management of water flow so as not to flood nesting areas, enables the plovers to nest. In addition, it is important that water not be drained away from nesting areas, because such drainage results in overgrowth of beach plants and destruction of open beach areas.

Piping Plover

Many of the wintering areas of the Red Knot are along coastlines, where increases in

Red Knot

sea level due to climate change are expected, so Red Knot population declines are anticipated. It is important to protect beaches used for Red Knot migration from human and dog disturbance and from removal of horseshoe crabs. Decreasing commercial harvest of gravid female horseshoe crabs will reduce the threat to Red Knots and mean that more crab eggs will be available to the knots. In addition, beach cleanups in such areas can help keep Red Knots and other seabirds from becoming entangled in plastic debris such as monofilament fishing line. Reduction of oil pollution on beaches and cleanup of oil spills are also important ways to reduce Red Knot deaths.

### GRASSLANDS AND PRAIRIES

Many of the bird species that rely on grassland and prairie habitat are being threatened. Some of these species are among the "in-trouble" birds discussed in this book. The main problems for many of them, including Gunnison Sage-Grouse and Greater Sage-Grouse, are overgrazing of brushlands, suburban sprawl, and construction of roads, power transmission facilities, and wind turbines, all of which are destructive to the sagebrush habitat required by these two species. These problems have resulted in reduction of the suitable habitat by over a third.

While there are many more Greater Sage-Grouse than Gunnison Sage-Grouse, both species face similar problems, such as the planting of large areas in crops (e.g., wheat or potatoes), overgrazing, and urban development. When there is human activity too close to the nesting area, sage-grouse will

Gunnison Sage-Grouse

*LEB*

often abandon their nests, especially if the disturbance is early in the incubation period. It is also important to manage burning of sage-grouse habitat. Prevention or reduction of fires often results in habitats being overgrown with larger plants less favorable for the sage-grouse, and too much fire can result in invasion by plants other than sagebrush.

The substantial decrease in the number of Greater Prairie-Chickens is due in large part to the loss of 80 to 99 percent of their prairie habitat. For example, in Iowa less than 0.1 percent of the tallgrass prairie that existed in 1830 still remains. While the prairie-chickens' original habitat was bison-grazed prairie, they can still use farmlands with remnant prairie, pasture, and hay and grain fields. Like other grassland species, prairie-chickens face many threats. These threats include agricultural practices such as grass removal and cultivation, use of insecticides that kill the birds' food supply, intense crop rotations that prevent growth of native grasses, and invasion of undesirable grasses. In addition, allowing brush and trees to grow in traditional prairie areas and not allowing lands to be periodically burned can be harmful to prairie-chicken habitat. In areas where tallgrass prairie segments remain, such as the Nebraska Sandhills and the Flint Hills area of Kansas, it is important that the habitat be conserved, preserved, and managed appropriately, which can be a difficult endeavor. Additional research is needed to determine the best balance of "benign neglect" of grasslands and careful interference to regulate plant growth to provide the maximum benefit to native prairie birds such as Greater Prairie-Chickens.

Much of the typical Lesser Prairie-Chicken habitat, which is primarily sand sagebrush and shinnery oak rangelands, has been altered or lost. The species requires a sand-sage habitat, but much of it has been converted for agricultural crop production and oil drilling, with oil industry infrastructure

Greater Sage-Grouse

and road construction also fragmenting habitat area. In addition, the extensive development of wind farms has caused problems because Lesser Prairie-Chickens abandon breeding grounds where wind farms have been installed. There is also evidence that some landowners eager for wind farm income may be less than eager to report the presence of Lesser Prairie-Chickens on potential wind farm sites, resulting in destruction or disturbance of the breeding lek areas. In some Lesser Prairie-Chicken habitats, such as the Texas Panhandle, oil shale extraction efforts cause fragmentation of habitat, which causes a reduction in Lesser Prairie-Chicken numbers. The lack of suitable habitat for Lesser Prairie-Chickens has caused their populations to be highly scattered and spread out in small areas of Colorado, New Mexico, Texas, Oklahoma, and Kansas. More than 70 percent of the areas where they remain today are privately owned and thus even more susceptible to human disturbance.

Many raptors, such as Ferruginous Hawks, also depend on prairie habitats for prey and nesting sites. It is beneficial to Ferruginous Hawks, both in winter and during breeding season, to have large ranch areas

Greater Praire-Chicken

Lesser Prairie-Chicken

that are not planted in crops. In areas that no longer have trees, construction of nesting platforms can reopen areas for breeding and thus increase the Ferruginous Hawks' nesting success.

When Mountain Plovers breed on agricultural lands, their nests are often destroyed by farm operations such as mowing or plowing, even though the birds themselves are quite tolerant of farm machinery and other vehicles. When they nest on recently mowed or plowed areas, they abandon the nests when the grass or crops, such as sunflowers or corn, grow too tall. Removal of native grassland animals such as prairie dogs and bison has resulted in a reduction in the numbers of Mountain Plovers nesting in these areas. Mountain Plovers will nest and winter on recently burned lands, and in fact controlled burning has been used to attract Mountain Plovers to western grasslands. People owning land in the Great Plains can help the Mountain Plover by managing their land to maintain or restore the short-grass prairie. Organizations in Colorado, for example, are working with landowners to provide education about Mountain Plovers and to help the landowners manage their property and flag nests of Mountain Plovers so that they are not destroyed.

Although Long-billed Curlews generally winter

Ferruginous Hawk

Mountain Plover

on or near southern beaches, they breed in grasslands. In most of their traditional breeding areas, the numbers of Long-billed Curlews are declining. Long-billed Curlews require that water be near their breeding grounds, even though they breed in upland areas. Many of their traditional breeding areas have been drained or the water diverted. Planting and invasion of the grasslands by tall stands of vegetation can also render habitat unsuitable for nesting by Long-billed Curlews. Similarly, in many of their coastal wintering areas the tidal marshes have been destroyed by urban and agricultural expansion, as well as by vehicle traffic.

It is known that urbanization, road construction, commercial development, and agricultural use of short-grass areas along the migration routes and in the wintering areas of the Buff-breasted Sandpiper have reduced the habitat that the species requires; however, properly managed grazing operations can improve sandpiper habitats.

### NORTHERN WET AREAS

In the Arctic breeding areas of Buff-breasted Sandpipers, human disturbance may cause abandonment of nests and increased predation by foxes and ravens. There are also concerns that development and oil drilling in the Arctic breeding areas of the Buff-breasted Sandpiper may increase the number of predators and reduce available breeding habitat.

Hudsonian Godwits, which breed on Alaskan and Canadian wet sedge meadows, are potentially threatened on their

Long-billed Curlew

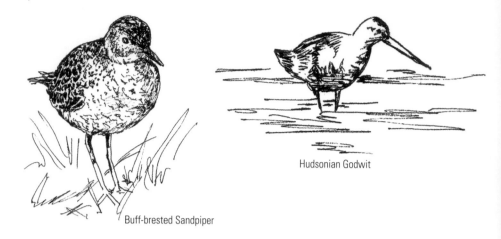

Hudsonian Godwit

Buff-brested Sandpiper

breeding grounds by development for oil and gas production, which disturbs and contaminates soil and water. Because of their small population size, any harm or disturbance to individuals of the species could tip the balance from stable to threatened.

### ISLAND AND OCEAN ENVIRONMENTS

Many of the bird species that breed on islands, such as a wide variety of seabirds, have had their numbers reduced due to habitat destruction. Due to the small numbers of the remaining Black-capped Petrels and Bermuda Petrels, which breed on Caribbean islands and islands in the Bermuda group, respectively, any additional destruction of habitat can be devastating to their populations as a whole. Such destruction can be caused by construction of buildings, logging, introduction of predators such as cats, dogs, pigs, or rats, or seawater pollution by oil or mercury.

Most of the nesting areas of Ashy Storm-Petrels are off the California coast, and they are already protected from destructive human activities, such as development, because they are within Channel Islands National Park and the Farallon National Wildlife and Wilderness Area. It is important that island protection be maintained and that predators on the nesting islands of the Ashy Storm-Petrels be eradicated.

Habitat loss is a major problem for Kittlitz's Murrelets, since they usually feed in glacial waters, and tidewater glaciers are retreating. Recreational and

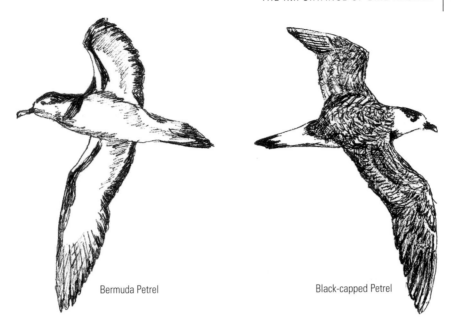

Bermuda Petrel                     Black-capped Petrel

Ashy Storm-Petrel

tour boats also disturb these murrelets, causing them to fly away from their feeding spots. There are also known deaths from oil spills; an estimated 7 to 15 percent of the Prince William Sound population was lost after the oil spill from the *Exxon Valdez* in 1989. It is believed that global warming, resulting in increased glacial melting, is the primary cause of their decline, since the decline is occurring even in areas where there is minimal human disturbance. As glaciers melt more rapidly and recede, there is less biological productivity in the waters and increased sedimentation, which makes it difficult for the murrelets to see prey in the water.

Most of the islands where Xantus's Murrelets (taxonomically split in 2012 into two species: Scripps's Murrelet and Guadalupe Murrelet) have traditionally bred had no native mammals. Introduction of mammals that eat these murrelets or their eggs or otherwise disturb the habitat or the nesting birds causes reduction in their numbers or extirpation of island colonies. Because

Kittlitz's Murrelet

Xantus's Murrelet

Island Scrub-Jay

these murrelets nest in colonies, any harm to an island, such as introduction of predators (cats, rats, mice, dogs) or other invasive animals (goats, sheep, rabbits), can cause that colony to go extinct, which has occurred on some islands. Eradication of introduced predators from nesting islands is critical. It is particularly critical to remove cats from Guadalupe Island, which is large and has been declared a biosphere reserve by the Mexican government.

Since 2002, when black rats were removed from Santa Barbara Island off California's coast, the Scripps's Murrelets have had greater breeding and nesting success rates. It is also important to control human access to areas conducive to murrelet nesting. Oil pollution from offshore wells or tankers is a serious threat to these two species of murrelets, as are gill nets and disturbance of nest sites. The number of these murrelets has decreased over the past century as they have been eliminated from some of their nesting islands. There is concern that a single Pacific oil spill could wipe out the Channel Islands breeding birds.

Because the Island Scrub-Jay lives on only one island—Santa Cruz Island off the California coast—there is concern that any type of disaster, such as disease or environmental destruction caused by introduced animals or development, could cause problems for the Island Scrub-Jay.

### FORESTS AND WOODED AREAS

It is believed that Flammulated Owls are declining in numbers primarily due to loss of habitat by logging in the western mountains, since their habitat appears to be restricted to forests of commercially valuable trees.

Flammulated Owl
LCB

Spotted Owls require dense wooded areas of mature and old-growth trees, which include many logs, snags, and live trees with broken tops. There is constant pressure to allow logging, development, and other uses of lands that are critical to Spotted Owls. In 2012 a final rule of the US Fish and Wildlife Service designating critical habitat for Spotted Owls was published. An objective of identifying critical habitat for Spotted Owls is to ensure that there is sufficient habitat, with the habitat features required by the owls, to support stable, healthy populations across their range.

Nesting habitat management and preservation in wooded areas having longleaf pines and other southern pines is critical to Red-cockaded Woodpecker population survival and growth. Controlled burning is needed to reduce

LEB

Spotted Owl

Red-cockaded Woodpecker

growth of deciduous saplings and brush beneath the pine trees. However, the waxy nesting trees of Red-cockaded Woodpeckers can be highly flammable, so particular care must be exercised in controlled burning. Allowing older, larger-diameter pine trees to remain standing when limited logging, particularly of deciduous trees, is taking place can be beneficial to these woodpeckers. Planting longleaf pine trees to establish new territories for potential nesting and to replace dying trees can provide habitat for Red-cockaded Woodpeckers.

The plight of the endangered Black-capped Vireo is due to their very limited range, their need for low shrubs or trees for nesting, and the fact that their preferred habitat is often in areas grazed intensively by livestock and deer, which leaves it unsuitable as vireo habitat. In addition, in areas where trees have grown to maturity and fires have not been allowed to burn, the land does not support the growth of the low woody plants that provide acceptable habitats for Black-capped Vireos. At the opposite extreme, agricultural activity and urban development have resulted in vegetation being removed from areas previously used by Black-capped Vireos, rendering the habitat no longer usable by these birds. In order to maintain and increase the number of Black-capped Vireos, their nesting areas must be managed to control plant growth. Prescribed burns and other actions to prevent growth of trees beyond a scrubby stage can help preserve

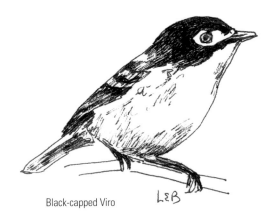

Black-capped Viro

LEB

acceptable habitat for the vireos. In addition, cowbird removal programs where Black-capped Vireos are found can lead to an increase in vireo nesting success.

Before the 1900s, Florida had many areas of scrub vegetation that were prime habitat for the Florida Scrub-Jay.

Florida Scrub-Jay

Many of these areas now feature airfields, cleared pastures, mines, pine and citrus plantations, highways, malls, mobile home parks, golf courses, and theme parks. These types of development eliminated the native saplings and shrubs that served as places for the jays to feed and nest.

The breeding area of Bicknell's Thrush is limited to high, wooded elevations in the northeastern United States and Canada. Support for conservation of balsam fir forest habitats in the northeastern United States and Canada is essential if there is to be a chance for Bicknell's Thrushes to maintain, if not increase, their population size.

While the Colima Warbler, which nests in wooded areas in the Chisos Mountains of southwestern Texas, is an endangered species, its limited habitat does not appear to be decreasing, primarily because areas where Colima Warblers typically breed in the United States are remote from human settlement. Their breeding areas in Big

Bicknell's Thrush

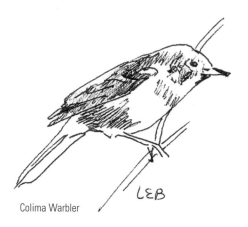

Colima Warbler

Bend National Park are protected and monitored. The main concern is that, with such a small suitable habitat, any human development, increased logging or agriculture, pollution, or particularly the presence of feral dogs and cats could threaten the population of Colima Warblers in habitats across the border in Mexico.

The Kirtland's Warbler requires jack pines for its nesting habitat and therefore breeds only in Michigan, Wisconsin, and Canada. While there are management practices in place for Kirtland's Warblers in these areas, there is continued need to monitor their populations.

The Cerulean Warbler faces habitat loss, as well as degradation and fragmentation of both its breeding range and wintering range. In its breeding range, the clearing of forest land for development and for agriculture continues to reduce the area where the Cerulean Warbler can breed, and, in its wintering range in South America, land is being cultivated for the production of coca leaves (for the illegal drug trade) and coffee beans. Although it is commonly believed that Cerulean Warblers require primary forest for their wintering

Kirtland's Warbler

habitat, this species is a perfect example of the importance of shade-grown coffee plants in the mountains of northern South America. While coffee monoculture in the tropical countries where coffee farms cover the landscape can be disastrous to wildlife,

Cerulean Warbler

the traditional, multilayered, shade-grown coffee plantations provide excellent habitat in which Cerulean Warblers and other migrant warblers can feed during their migration or spend the North American winter months.

Most of the problems facing Golden-cheeked Warblers, both on their breeding range and on their wintering range, are related to the growth of the human population. Because its nesting areas are in and near heavily populated areas of Texas, the Golden-cheeked Warbler faces serious habitat loss or fragmentation of habitat as a result of agriculture or the construction of reservoirs and subdivisions. It is estimated that between 1962 and 1974 the population of Golden-cheeked Warblers decreased by more than 10 percent. Wildfires at Fort Hood, home to a large percentage of the Golden-cheeked Warblers, destroyed about a quarter of the breeding habitat there in 1996. As the nesting areas for Golden-cheeked Warblers are fragmented by clearing or other

Golden-cheeked Warbler

disturbance, there is concern that there may be reduced genetic diversity in the isolated breeding populations, in addition to the loss of nesting sites and nests. The winter habitat of Golden-cheeked Warblers in Mexico and Central America is also in danger from lumbering, mining activities, the cutting of firewood, and the clearing of land for agriculture.

# WHAT CAN BE DONE TO HELP BIRDS

What can we do for an "in-trouble" bird? This section provides a starting place for you to begin learning about birds in danger and offers encouragement and help in finding ways to be part of the solution to the problems facing birds.

There are many ways we can help "in-trouble" birds, and some of these approaches relate to work being done by organizations that are devoted to such efforts. Although I did some research on the various organizations that focus on the protection and preservation of each species (and its habitat) covered in this book and have summarized the information that I found, I cannot personally vouch for all of these organizations nor can I guarantee that they will still be functional organizations or efforts by the time you read this book.

Most of the organizations that I have mentioned can use your financial help, and many of them would also welcome your becoming involved with their projects. Even learning more about these organizations can help you become an informed citizen who is better able to analyze and understand the environmental, legislative, and other issues that can affect birds and their habitats.

The narrative presented on each bird in trouble includes additional ways to help. I hope that this book stimulates you to choose a bird (or an organization) and learn more about it—and then do something to be part of the solution to the trouble facing that bird. Sometimes a particular bird species can "grab" us. Something about its plight, or its appearance, or maybe some intangible essence of the bird speaks to us. If you have that sort of reaction to one or two (or more) of the species covered in the book, I encourage you to learn *everything* you can about the bird species by going to the Internet and to the references found there as well as in the Suggested Reading section of this book. The more you learn, the better you can help make an informed evaluation of what can be done and what you can do to help a particular bird species.

Following, in no particular order, are things that you might consider doing to help birds in general.

## OBSERVING BIRDS

As you will note when you read the species accounts in this book, scientists and conservationists lack data on many bird species, so finding solutions to the problems facing birds is that much more difficult. Each of us who is a birder is encouraged, for example, by the National Audubon Society, to help support its efforts by providing information on our sightings to the appropriate organizations, some of which are mentioned in the species accounts. By reporting our sightings on eBird (http://www.ebird.com/), which is free via the Internet, we can contribute data on species population size, trends, and changes in range.

If you like to go out birding and can identify birds, you should consider participating in Breeding Bird Surveys (http://137.227.245.162/BBS/index.html) sponsored by the Patuxent Wildlife Research Center, operating under the US Geological Survey (http://www.pwrc.usgs.gov/), and Environment Canada's Canadian Wildlife Service (http://ec.gc.ca/reom-mbs/default.asp?lang=En&n=416B57CA-1). Whether you are an expert birder or not, you can join others in doing Christmas Bird Counts (CBCs, http://birds.audubon.org/christmas-bird-count). They are done all over the country and provide a wealth of information about changes and trends in bird numbers and distribution.

## LEARNING AND SPREADING THE WORD

Educating the public about birds can be as simple as telling someone else about what you learn from this book or from your further reading and research.

If you are already a birdwatcher, you can help others become aware of birds and more knowledgeable about them by donating bird books (e.g., field guides) to a local library or by assisting with field trips sponsored by local bird or nature organizations.

This book only touches the surface of information on in-trouble birds and on habitat concerns related to birds. You can become better informed about a particular problem habitat or in-trouble bird by starting with the Internet and the sources listed in the Suggested Reading section of this book. If you learn more about birds, you can then teach others, especially the young, how to identify bird species and about the value of birds, as well as how people can minimize damage to bird habitats.

You can also become aware of federal and state laws related to birds and their habitat. These laws are not static. There are proposals for strengthening them, or weakening them, being proposed and considered by legislative bodies all the time. Stay informed.

The more you learn, the more you will have information and understanding to assist you in informing lawmakers about issues that affect birds and the environment. Action on behalf of birds can be useful in such matters as promotion of sound agricultural, forest, and wetland policy, as well as helping support actions to fight climate change.

### HABITAT

One thing you might try to find out for a particular species is whether there is an Important Bird Area (IBA) designated for its current habitat. IBAs are typically established under the auspices of state Audubon chapters (http:// web4.audubon.org/bird/iba/). If such an area does not exist, you can do research to find out whether there is an area that should be listed as an IBA. For example, you could contact local bird organizations near where the bird species is found. By helping to gather more detailed information on the bird's presence in a specific area, you might possibly garner sufficient support to have a location designated an IBA.

When looking for birds in the open, stay on trails whenever possible to avoid harming birds and the environment that they need. Don't go into areas where birds are nesting, such as marked beach areas, during breeding and nesting season.

If you own land, consider the possibility of conservation easements, especially if you are in an area with habitat that might be used by an "in trouble" species.

For your local birds, you can put up birdhouses, birdbaths, and bird feeders. There are more details on these bird amenities in the "Helping Everyday Birds" section of this book.

Do not discard plastic fishing line into waterways, lakes, or ponds, on the beach, or anywhere else where birds can become entangled in it.

Cut up plastic that has any holes in it, such as plastic beverage bottle holders, into small pieces so that if the plastic comes into an area where there are birds, such as the beach or a landfill, the birds cannot become ensnared by having their feet, necks, or whole bodies caught in the holes.

Plant native plants for shelter near a birdbath to give birds a place to hide and roost.

Turn off lights at night. Nighttime migrants are attracted to lights and are killed by colliding with windows or light fixtures.

### ORGANIZATIONS

You can support bird life by joining and donating to a conservation organization; many such organizations are mentioned in this book. Organizations devoted to conservation and birds include the National Audubon Society (http://www.audubon.org/) and the American Bird Conservancy (http://www.abcbirds.org/).

Partners in Flight (PIF, at http://www.partnersinflight.org/), begun in 1990, is a response to growing concerns about declines in the populations of many land bird species, especially those for which there were no existing conservation initiatives. It is a cooperative effort involving partnerships among federal, state, and local government agencies, philanthropic foundations, professional organizations, conservation groups, industry, the academic community, and private individuals. It focuses primarily on helping species at risk, keeping common birds common, and establishing voluntary partnerships to bring together people who will want to support birds and their habitat. You can help birds by learning more about PIF and possibly becoming involved in its work.

Throughout this book I refer to organizations that are involved in helping birds in trouble or in protecting the environment. You can volunteer to help one of the conservation organizations mentioned in this book or one that you learn about through your future exploration of this topic. If you have the ability, consider supporting one or more of these organizations financially or with your time and labor. One organization that works to purchase and preserve land that is vital to various species is The Nature Conservancy (http://www.nature.org). Many states have local land conservancies with similar goals.

You might also support the Center for Biological Diversity (http://www.biologicaldiversity.org/), which is working to publicize the plight of endangered animals and to get in-trouble species considered for protection under the Endangered Species Act.

The Environmental Defense Fund (EDF, at http://www.edf.org/) works to preserve the natural systems on which all life depends, with emphases on climate solutions; ocean fishing reforms; creating economic demand for sustainable and climate-adaptive farming, ranching, and water efficiency programs; and reducing exposure to pollutants.

### OTHER ACTIONS

You could write a letter or, even better, letters or emails or place telephone calls to your elected officials to convey concerns about issues related to conservation and endangered species as you become more aware of and knowledgeable about the issues affecting birds.

More generally, you can help the environment, including birds and their habitats, by reusing and recycling materials.

Do you drink coffee? Try to buy shade-grown coffee, preferably varieties raised in a more traditional way that gives songbirds such as Golden-cheeked Warblers a place through which to migrate or to spend the winter in Central and South America and the Caribbean. The National Audubon Society lists the bird species found in shade-grown coffee plantations in Latin America at http://web4.audubon.org/bird/at_home/coffee/species/index.html.

If you are a hunter, know your waterfowl and other game birds, and know the laws and obey them!

Often there is not a single specific way to support a species that is in trouble, but there are many general things you can do, such as lobbying or working to conserve habitat needed by the species. Thus, in addition to working to improve the situation for a particular species of concern, if you are interested in a particular type of habitat, such as wetlands or prairie grasslands or forests or sagebrush deserts, you can contact organizations that have a broader scope of endeavor than conserving a single species. Of course, even work by the organizations dedicated to a particular species is likely to have a positive effect on other species that share the same habitat and are affected by the same problems (e.g., development).

Because I am a birder, birds are my window into nature. You may find that you are more passionate about other living things. Wonderful! There are other animals and plants that are also in trouble. Whatever you can do for those species is also of great importance. It usually comes down to making the environment a better place for living things, whatever they are.

This is not a comprehensive list of things you might do, but it can get you started. You can be creative. One of the most creative things that I have read about recently is the Species on the Edge Art & Essay Contest for children that was started in New Jersey to create awareness about endangered wildlife there. Not only is this a way to let young people begin to learn, but the winning artwork is displayed around the state and is also published in a calendar. Only if each generation is concerned about birds in trouble will there be any chance of making a change.

# SPECIES ACCOUNTS
# OF BIRDS IN TROUBLE

The accounts of these "in-trouble" birds are presented in the basic (and current) taxonomic order, that is, the accounts begin with those bird species that scientific evidence indicates are more genetically primitive and proceed in succession with species that are understood to be more genetically advanced. If you are seeking information about a particular bird but are unfamiliar with taxonomy, check the index. You can also become more familiar with the taxonomic organization of birds by looking at the website of the American Ornithologists' Union (AOU, at www.aou.org). Please be aware that the way bird species are named and classified regularly changes as new information becomes available about species.

For each of the bird species discussed here, the "Global Population Estimate" gives the best available data on the number of birds of that species as indicated in the National Audubon Society's Watchlist 2007 and also notes whether the global population is limited to the Americas. In some cases, while there is an officially recognized estimate for the number of birds of a species, different recently published accounts provide other estimates. Appendix 1 also presents the estimates from the Audubon Watchlist 2007, as well as other estimates, if available, where the literature seemed to indicate some serious uncertainty about the actual number of birds of a particular species. In very few species is it possible to know exactly how many birds there are at any one time, and therefore these numbers are provided only to give a basic idea of the species' status.

For each species, a brief description, supplemented by paintings, is provided to indicate what the bird looks like, how it behaves, what type of habitat it prefers or needs, and where it tends to breed and to winter (all of the birds described here spend at least part of the year in the United States or on offshore waters of the United States). This section is not intended to serve as an identification guide but rather is designed to provide basic species information to those who are not familiar with the species. Much more detailed information about each bird species can be found in bird books and on the Internet. Particularly informative Internet sites include several sponsored by

the Cornell Lab of Ornithology: http://www.allaboutbirds.org/, http://bna
.birds.cornell.edu/bna/ (requires a subscription), http://neotropical.birds.
cornell.edu/portal/home, and http://www.stateofthebirds.org/. Also valuable
are the various updates of the Audubon Watchlist (e.g., http://birds.audubon.
org/species-by-program/watchlist) and even Wikipedia (http://en.wikipedia.
org/wiki/).

There are also numerous bird-identification apps that can be downloaded
to your cell phone and electronic reader. Examples include iBird Pro: North
America, National Geographic Birds, Audubon Birds, and the Sibley eGuide
to the Birds of North America.

Information about the status of each bird species from the period of
the early US naturalists such as John James Audubon (1785–1851) is summarized in the account of that species. For some species, very little or nothing is known about their status during that period, but whatever is known
or surmised about the population size before the twenty-first century is
summarized for each species covered. In a few cases I could find very little
early information, but that does not mean such information is not available
somewhere. I welcome any leads to information (about this as well as about
other facts related to these species) I may have missed so that I can be better
informed myself about these birds.

After presenting information on the bird species' early status, each species
section explains what factors are known or believed to be causing problems
such as a decline in numbers or a steady but dangerously low population. In a
few cases where the population size of the bird does not seem to be decreasing or to be in imminent danger of decreasing, any other reasons for the
troubled status of the species are provided.

While general information on how to help birds is provided elsewhere
in this book, information about helping particular species, to the extent
that there are organizations concerned with helping a particular species, or
particular actions that can be taken to help that species may be found in the
section on that species. Such details include information on what is currently
being done for the species, as well as what actions are believed necessary or
desirable to improve the status of the species. If you are interested in doing
something for a particular bird species, you might begin by reviewing the
organizations' websites and then contacting people at the organization for
further details of what it does and whether financial support for that organization, or working with and for the organization, might be of benefit to a bird

in trouble. Even if there is not a particular thing you can do directly for the bird species, following up on the account in this book by going to listed websites and organizations can help you become better informed and better able to evaluate what others are saying and the actions that are being proposed or should be proposed.

The discussion of each species concludes with "A Personal Note," which provides a few highlights of my experience with each species. The reason for this section is to help you gain a small understanding of what it is like to look for and find the in-trouble birds that are covered in this book. In some cases, my experience is lamentably minuscule, particularly with the rarest of the species covered in this book, even though I have been birding for many decades and have been trying to observe all of the birds of the United States and Canada. This section is of course biased by where I have lived and birded. My most active birding years have been those since 2000, when I lived in Texas (2000–2011), South Dakota (2011–14), and Alaska (2014–present). Therefore, these personal notes for birds that are found in these three states include primarily accounts of my seeking and finding these species there. A number of other personal notes about birds not found in Texas are from 2008, when I birded extensively all across the United States.

Since I was not personally familiar with some details on the status and preferred habitats of many of the species covered in this book, I have consulted a wide variety of sources to learn as much as I could about them. Thus, much of the factual and historical information on the status of the bird species discussed in this book was gleaned and compiled from bird books and other books, as well as from general websites and specific websites about particular species and habitats, and is believed to be accurate. I am grateful that such information is available, and I hope that you will expand your search for information beyond what is presented here, which is of necessity condensed.

# Trumpeter Swan (*Cygnus buccinator*)

Global Population Estimate: 34,803, all in North America

Although most people are familiar with the long white neck and silhouette of swans, it is often a shock to realize how big these birds are. Trumpeter Swans have a five-to-six-foot wingspan. Their vocalizations sound very much like a loud car horn as they call to other swans or attack an invader in their territory. While swans' long necks give them a graceful appearance as they float on the water, when swans waddle on land they have a clumsy, awkward appearance. When they fly, their long white necks stretch out ahead of their white bodies and their black feet extend backward beyond their tails, while their wings beat slowly and regularly. The young birds are brownish-gray for their first year.

Trumpeter Swans feed by plucking vegetation either on land or in the water while they swim, tipping their bodies so their heads are underwater and their rear ends are all that can be seen, similar to puddle ducks such as the Mallard. They eat plant roots, leaves, seeds, aquatic insects, land snails, small reptiles, and other small animals.

The majority of Trumpeter Swans breed in central Alaska and western Canada and winter on the western Canadian coast. There are also year-round populations in the United States, in Washington, Oregon, Idaho, Montana,

Minnesota, and Wisconsin. Some Trumpeter Swans wander even farther south and winter in California and some of the interior western states.

In spring, most Trumpeter Swans head north to nest in large remote areas with rivers, lakes, or marshes. After the young are raised, the swan families head back south to another wetland area.

Trumpeter Swans were once widespread across Canada and the northern United States, being particularly common in the West. Because of their similarity to the more common Tundra Swans, it is unknown how many there were early in US history. In the journals of Lewis and Clark, written during their journey of exploration across the West, is a statement that "the Swans are of two kinds, the large and small. The large Swan is the same with the common one in the Atlantic States. The small differs from the large only in size and note; it is about one fourth less, and its note is entirely different. These birds were first found below the great narrows of the Columbia, near the Chilluckittequaw nation. They are very abundant in this neighbourhood, and remained with the party all winter, and in number they exceed those of the larger species in the proportion of five to one."

As settlers moved west, Trumpeter Swans were reduced in number or eliminated from many areas. Loss of suitable swan habitat due to human activities began in the early 1600s. The dramatic decrease in numbers and extirpation from some areas were also partly due to the demand for swan feathers, used as quill pens, and for skins, as well as because the young swans (cygnets) were considered tasty.

As a result of unregulated hunting, Trumpeter Swans became nearly extinct in the United States, with only a few thousand remaining at the beginning of the twentieth century. In the 1930s and 1940s, naturalists became concerned that the Trumpeter Swans were likely to become extinct. By the 1930s there were only sixty-nine birds south of Canada. Even though Trumpeter Swans have been reintroduced in many areas since then, they still numbered only about thirty-five thousand as of 2005. While this figure may seem like many birds, it is a small enough population that a disease, a bad winter, loss of habitat, or polluted waters could cause the numbers to head back toward extinction.

Trumpeter Swans have been federally protected under the Migratory Bird Treaty Act since 1918 and can no longer be legally hunted. It would seem that birds such as Trumpeter Swans, which do not usually nest near humans and can be very vicious if humans do come near, would not be affected adversely by humans if hunting were strictly controlled or eliminated. Hunting, however, has been a problem and still does cause reduction in numbers; the beauty and size of Trumpeter Swans makes hunters value their feathers, skins, and meat. In addition to the fact that illegal hunting still occurs, there is also the mistaken killing of Trumpeter Swans by people who are legally hunting the more common Tundra Swan.

When there are human activities such as boating, birdwatching, or photography near Trumpeter Swan nesting areas, the swans are prone to abandoning their nests or leaving their flightless young undefended against predators. In the wintering areas, human disturbance can cause Trumpeter Swans to expend too much energy, resulting in weakness, reduced capability of reproduction the following breeding season, or death.

Swans are very sensitive to lead and are regularly poisoned by ingestion of lead shells or lead shot. Spent ammunition remains in the bottom sediment of many lakes that have been hunted for waterfowl over the years, even though there has been no legal use of lead shot for a long time in many locations.

While many of these problems still do take a toll on Trumpeter Swan numbers, unlike with many of the other species discussed in this book there are many management and conservation efforts that have resulted in increases in the numbers and distribution of Trumpeter Swans.

Early efforts to help Trumpeter Swans included outlawing shooting, reducing predators, feeding the swans in the winter, and moving some of the swans to better areas for breeding.

There have been many Trumpeter Swan reintroduction programs in the northern United States, which in some areas are resulting in a turnaround in Trumpeter Swan population numbers. Importantly, these programs have emphasized restoring Trumpeter Swans to their former breeding areas, such as those in Wisconsin, Michigan, Ohio, Iowa, Arkansas, and Ontario. Efforts to try to introduce Trumpeter Swans to other areas, such as more urban areas where Trumpeter Swans are not native, have often resulted in problems such as the introduced swans chasing other wildlife out of the swans' new homes and threats by the very territorial swans to people who venture into the swans' territories.

There are three major federal plans for managing Trumpeter Swans that follow major migratory flyways—the Mississippi and Central Flyway Management Plan for the Interior Population of Trumpeter Swans, the Pacific Flyway Management Plan for the Rocky Mountain Population of Trumpeter

Swans, and the Pacific Flyway Management Plan for the Pacific Coast Population of Trumpeter Swans—as well as a number of state management plans.

It is interesting to note that a publication from 2012 summarizing research by the National Park Service indicates that the trend toward warmer, longer summers in Alaska and the slowed onset of winter has led to some expansion of the range of Trumpeter Swans in Alaska because of the apparent direct link between temperature and the presence of breeding Trumpeter Swans in Alaska.

The Trumpeter Swan Society (http://www.trumpeterswansociety.org) is a nonprofit organization dedicated to ensuring the vitality and welfare of wild Trumpeter Swan populations. The website indicates that the group "is a means for those who cherish these swans to work together and support their continued restoration." In addition to supporting the work of this organization, if you live in an area where Trumpeter Swans occur and would like to explore the potential for your wetland, or another wetland, to support successful swan nesting, you can find helpful information from the Trumpeter Swan Society. You can also "adopt a swan" and volunteer with the society in many different ways.

A Personal Note: Trumpeter Swans, and their more common, smaller relative, the Tundra Swans, are rare in most of the country. Most birders, whether seasoned expert or beginner, welcome the chance to see Trumpeter Swans. Texas birders (of which I was one when I began work on this book) can mostly only dream about Trumpeter Swans, unless one heads to the far northern United States or Canada. When someone reports them in Texas, such as a couple of swans that were seen in 2003 in the northeastern Texas Panhandle, birders often drop everything and drive to the site to check them out, as I did. A couple of years later, I had the good fortune of seeing a family of Trumpeter Swans on the ponds of a large Panhandle ranch. After a friend and I bumped along a ranch road, checking out pond after pond from a distance, trying not to spook whatever birds were on the ponds, we finally spotted a huge pond in the distance with what seemed to be whitish dots among smaller darker dots. Only after we crept up toward the pond, shielded by low brush, could we see the five Trumpeter Swans, slowly swimming along among the ducks and majestically dipping their heads down to nibble at things in the water—two elegant white parents and their three grayish youngsters.

In September 2014, when my husband and I moved to Alaska, I was delighted to find pairs of Trumpeter Swans floating in many roadside wet areas on our route from Yukon Territory into Alaska. I saw more Trumpeter Swans in and around Anchorage in the next month or so as swans made their way southward on migration, and in fact they were much more common than the Tundra Swans then seen in the open waters around Anchorage.

---

GUNNISON SAGE-GROUSE AND GREATER SAGE-GROUSE

Gunnison Sage-Grouse and Greater Sage-Grouse are very similar large, heavy grouse with pointed tails and are found in dry sagebrush areas. The males display for the females in spring in the western North American plains, often in the mountains. Although these two species are very similar, their ranges do not overlap. Similar to a number of other chicken-related birds, both sage-grouse species display in leks, which are generally house-sized or larger areas where the males gather each day during breeding season, with each male working to hold on to and enlarge his territory within the lek by fighting with or displaying before the other males in order to become the dominant male and attract more female breeding partners.

### Gunnison Sage-Grouse (*Centrocercus minimus*)
Global Population Estimate: 2,000 to 5,000, all in North America

The primary place that Gunnison Sage-Grouse may be found is near Gunnison, Colorado. As of 2006, there were seven populations—six in Colorado and one in both Colorado and Utah. The largest population is in the Gunnison Basin.

The Gunnison Sage-Grouse, like the Greater Sage-Grouse, is big and brown with a black belly. The male has a white breast that expands and contracts during display while he makes a series of low hoots. The male has a clump of thick plumes on the back of his head that are raised when he displays. The female's breast is brownish, and she does not have the head plumes.

Because Gunnison Sage-Grouse was not originally defined as a separate species but was typically grouped together with Greater Sage-Grouse, the size of the Gunnison Sage-Grouse population in historic times is basically unknown. It is believed that the population was at least several times greater than it is today and occurred in an area limited to southwestern Colorado, northwestern New Mexico, northeastern Arizona, and southwestern Utah, an area of 21,370 square miles.

While the Gunnison Sage-Grouse is not federally listed as endangered,

both BirdLife International and the National Audubon Society consider it endangered. It is estimated that the population of Gunnison Sage-Grouse has decreased by 66 percent since 1953, with some of the populations being extirpated since 1985 and others decreasing precipitously. The current amount of land occupied by Gunnison Sage-Grouse is estimated at 1,820 square miles. Recently, some years of heavier rainfall have resulted in an increase in numbers at some locales. It is estimated that five hundred to fifteen hundred birds are needed to maintain each population in a stable condition so that susceptibility to disease or inbreeding problems do not jeopardize the species.

On April 12, 2006, the US Fish and Wildlife Service announced that it would not add the Gunnison Sage-Grouse to the federal list of threatened or endangered species. The FWS also removed the species from the candidate list (to which it had been added in 2000). Local conservation plans have been approved for six of the seven populations. There are numerous landowners who have expressed interest in voluntary conservation efforts designed to prevent the need for federally listing the Gunnison Sage-Grouse and to allow the landowners to use their lands.

As of May 2010, the Colorado Department of Natural Resources had a

Gunnison Sage-Grouse Rangewide Conservation Plan (http://wildlife.state
.co.us) in place, one designed to provide the latest in scientific knowledge
with respect to minimum viable population size and habitat requirements,
with the purpose of protecting, enhancing, and conserving Gunnison Sage-
Grouse populations and their habitats. This plan is deemed necessary for
consistent and timely habitat improvements and population expansions so
that the species can be removed from consideration by the FWS for endangered status.

In 2013 the FWS proposed to list the Gunnison Sage-Grouse as endangered, but in April of that year Colorado Parks and Wildlife responded that
the listing was not needed due to aggressive conservation actions by the state
and counties and the fact that the populations of this species were stable and
at "historic highs." After a fight of nearly two decades, in November 2014 the
FWS ruled that the Gunnison Sage-Grouse was "threatened" and not "endangered." Although this designation appears to be a final decision, it is being
challenged by communities in Colorado, and riders in federal legislation have
delayed implementation for at least a year.

There are state and local programs that have been established for controlling predators, reseeding sagebrush, and limiting human access in some
areas. There also have been and continue to be efforts to move sage-grouse to
areas with better habitat or to improve the genetic diversity in areas where the
sage-grouse populations are isolated. In some cases they have been removed
from areas where they have been damaging cropland.

To better inform the public about the Gunnison Sage-Grouse and its status (and allow birdwatchers to see this very rare species), there is a publicly
viewable lek east of Gunnison, Colorado. There are also school programs
and brochures. Conservation groups have worked to protect the Gunnison
Sage-Grouse locally, but a substantial portion of the range of the grouse is not
covered by such protection.

While there is little that most individuals can do directly to help the Gunnison Sage-Grouse, learning about Colorado's Gunnison Sage-Grouse Range-
wide Conservation Plan is a good first step. If you are (or know) a private
landowner near Gunnison or another of the areas inhabited by Gunnison
Sage-Grouse, you can participate in, or encourage, better land-use practices
in grazing, mining, and energy extraction, as well as habitat reconstruction to
link the various populations.

You also can join the High Country Citizens' Alliance (http://www

.hccaonline.org/page.cfm?pageid=1951). The mission of this alliance is to champion the protection, conservation, and preservation of the natural eco-systems within the Upper Gunnison River Basin. This alliance is a grassroots organization with goals focusing on the health and biodiversity of the local environment—land, water, air, and wildlife. The group collaborates with interested and affected parties to reduce global climate change and protect public lands, rangelands, water resources, and endangered species, specifi-cally including the Gunnison Sage-Grouse.

A Personal Note: The Gunnison Sage-Grouse had not even been defined as a separate species when I first birded in Colorado. Later, in April 2008, I took a birding trip to Colorado. I scheduled the predawn time of my first morn-ing in Colorado at a blind near Gunnison, since Gunnison Sage-Grouse, like other species in which the males gather to display for the females, only do so in the very early hours at sunrise. Unfortunately, due to heavy snows the day before, no grouse appeared at all at the blind, probably because they were dozing somewhere in a cozy snow bed. Having a tight schedule that allowed time to see other Colorado birds, I could not stay around the next morning to try again for the Gunnison Sage-Grouse. In spite of the snows, however, I was easily able to find the more common and widespread Greater Sage-Grouse (see below) as it hunkered down to try to escape the cold wind. I had left my schedule empty for my last day in Colorado on this trip, so I took the opportunity to go back to Gunnison for one more attempt. Again, long before dawn, I drove to the blind and parked, aiming my car toward where I hoped the grouse would appear (other cars were similarly parked—you are not allowed to get out of your cars while there). It was reminiscent of being at a drive-in theater, but silent and very cold and of course without a screen. Although still snowy, the ground did have clear areas where the snow had blown or melted away since the big storm. The grouse, eager to get on with their mating ritual, did appear. More than thirty of them were spread out across the generally flat land at the base of the nearby hill, males with their stiff, pointed, fanned tails, puffing out their gular (throat-area) patches in spastic spurts, and females casually pecking at the earth and seeming to ignore all the posturing and excitement around them. As the day became lighter, the grouse departed, one by one, flying away over the surrounding low hills and leaving the area barren, as if no grouse had ever danced there.

## Greater Sage-Grouse (*Centrocercus urophasianus*)

Global Population Estimate: 150,000 or fewer in the United States (eleven states) and Canada (two provinces)

Like the Gunnison Sage-Grouse, the Greater Sage-Grouse is best known for the males' elaborate courtship displays in sagebrush habitats in western North America. In the courtship displays, the Greater Sage-Grouse male erects the feather plumes on his head and spreads his tail wide, with the pointed tips of the feathers individually visible. He rapidly inflates and deflates yellow air sacs on his neck so that the large white ruff on his neck and breast that overlies the air sacs flops up and down, making a sound that has been described as a "loud, bubbling popping," while the plainer females peck at the ground with apparent nonchalance. Periodically the males engage in roosterlike face-offs with other males gathered on the leks. Eventually, the females acquiesce to mating, usually with a small number of the displaying males. The only somewhat similar species to the two sage-grouse species is the Sharp-tailed Grouse, which shares habitat with Greater Sage-Grouse and faces similar problems, but it is a smaller bird, with a tail that is largely white rather than dark.

The diet of the Greater Sage-Grouse is primarily sagebrush, supplemented in summer by insects such as grasshoppers, beetles, and ants.

Before regulations were in place, commercial hunting of Greater Sage-Grouse caused a decline in their numbers, and even regulated hunting has been estimated to have reduced the population by 5 to 15 percent or more in some areas. In fact, by 1870, there was sufficient concern about sage-grouse numbers to pass legislation in Montana (and in Colorado by 1877) so that hunting of these birds was regulated to prevent excessive harvest.

Greater Sage-Grouse are essentially limited to sagebrush habitats, which occur only in the western United States and Canada and were once widespread over much of western North America (sixteen states and three provinces). In direct proportion to the dramatic decrease in sagebrush habitat, the range of Greater Sage-Grouse has also dramatically decreased. Today they are found primarily in eastern Montana, much of Wyoming, northwestern Colorado, Utah, the southern half of Idaho, the northern two-thirds of Nevada, northeastern California and along the eastern state line, and in southeastern Oregon, with smaller populations in the western portions of the Dakotas and in central Washington. They have been extirpated from British Columbia, Arizona, New Mexico, Oklahoma, Kansas, and Nebraska.

Pesticide use can reduce their insect food supply, and herbicides may reduce the cover needed by the sage-grouse. It is unclear whether hunting or natural causes, such as severe winter weather, yield greater reductions in Greater Sage-Grouse populations. In any case, many people are probably more concerned about the status of Greater Sage-Grouse than they are about other birds, because it is a game bird. Because of the remote and vast areas inhabited by Greater Sage-Grouse, in some areas it is unclear how bad the problem is.

Maintenance and improvement of habitats for Greater Sage-Grouse are very important for managing their numbers, and therefore specific measures are recommended with respect to grazing levels, controlled burns, and restoration of the habitat, particularly in nesting areas. The National Audubon Society Important Bird Area program helps to conserve Greater Sage-Grouse. In addition, volunteers can help monitor the species population by participating in the relevant Christmas Bird Counts. National wildlife refuges are important in providing habitat for Greater Sage-Grouse, and it is therefore critical that they not be underfunded.

The Western States Sage and Columbian Sharp-tailed Grouse Workshop, part of the Western Association of Fish and Wildlife Agencies (http://www

.wafwa.org/), meeting in even-numbered years, is working toward collection and sharing of data among states and developing management plans for the particular grouse populations in those states. Management recommendations and guidelines are set forth in a document from the Western States Sage and Columbian Sharp-tailed Grouse Technical Committee.

Creative financing programs are raising money for conservation of sagebrush-dependent species such as the Greater Sage-Grouse under the auspices of organizations as diverse as the Oregon Cattlemen's Association (which is backing legislation to place a tax on salt) and western bird groups (which are advocating for a tax on bird seed).

The FWS, which designated the Gunnison Sage-Grouse "threatened" in 2014, is expected to decide on the status of Greater Sage-Grouse late in 2015; it is now listed as a "candidate" species.

A Personal Note: In 1996 I saw my "lifer" sage-grouse in Colorado. At that time, as far as I was aware, there was only one sage-grouse species. By 2008, when I returned to Colorado to bird, sage-grouse had been split into Greater Sage-Grouse and Gunnison Sage-Grouse. Because the latter has such a tiny range (see above), I assume that the bird I had seen earlier was the more common Greater Sage-Grouse. In April 2008, I remedied any uncertainty, however, and saw many of each species. At first, though, due to a snowstorm dumping more than three feet of snow across most of the Colorado mountains, I was not even certain that I would be able to travel to where the Greater Sage-Grouse can usually be found. After a couple of days in Colorado, I reached Walden, which is in an area that was supposed to be good for Greater Sage-Grouse. About ten to twelve miles south of Walden, I noticed dark bird-blobs ahead among some stunted sage. The blobs turned out to be Greater Sage-Grouse, twenty-two of them hunkered down, all facing into the wind, most of them sitting behind individual sage clumps and weathering it out. I drove farther south and found ten more. The next day, near Coalmont, as I drove into another area where the Greater Sage-Grouse were supposed to be, a Greater Sage-Grouse crossed the road ahead of me, and then I saw about twelve more birds in the snow, coming up onto the road, probably to get out of the deep snow that was making it difficult for them to walk. Then all of them started walking up the road toward me, mostly pecking at the vegetation. Only one sort of puffed up for a while and looked like he wanted to display, but he apparently changed his mind, even though it was the season for mating.

GREATER PRAIRIE-CHICKEN AND LESSER PRAIRIE-CHICKEN

Both prairie-chicken species are famous for their courtship behavior in which the adult males gather on leks (for prairie-chickens, these are known as "booming grounds"), with each male defending his own territory. The male spreads his wings so that they are bowed out and puts his head down toward the ground, stamps his feet, and makes a booming sound. The male has orange combs over his eyes and elongated neck feathers that he raises during displays. When the neck feathers are erect, the male's inflated gold neck sacs are visible. Females join the males at the lek, and at some point after the displaying has gone on for a while they mate with the dominant males.

## Greater Prairie-Chicken *(Tympanuchus cupido)*

Global Population Estimate: 690,000, all in North America

The Greater Prairie-Chicken is a round, chunky, chicken-sized bird with brown, uniformly barred feathers and a short, rounded tail. The typical diet of a Greater Prairic-Chicken includes the seeds of grasses, corn, and fringed sage, as well as dandelion and clover and tree buds.

Greater Prairie-Chickens need both heavy grass prairies in which to roost

and nest and open areas for their courtship displays. Thus, it is important that they have large, open areas of mid- and tallgrass prairie with few trees and that there are elevated booming areas with short and sparse vegetation, as well as denser areas where the nests can be placed. They do not migrate and usually remain in the same area year round, although a small number of them make forays away from the breeding area after breeding.

The population of Greater Prairie-Chickens has decreased by about 97 percent since the early 1800s. When Europeans came to this continent, there were three subspecies of Greater Prairie-Chicken, but since then the Heath Hen has gone extinct through loss of prairie habitat and the endangered Attwater's Prairie-Chicken, which was similarly reduced in numbers because of grazing pressure and increased cropland, is being kept alive through breeding in captivity and repeated releases, with only about fifty or fewer birds remaining in the wild. The remaining subspecies, while extirpated from many states and Canada, still remains in ten prairie states. The main impact on the prairie-chicken arose in the nineteenth century, when the Greater Prairie-Chicken faced two serious problems: heavy hunting and loss of prairie habitat. For example, there are accounts of three hundred thousand Greater Prairie-Chickens being shipped to market from areas of Nebraska in 1874, and twenty thousand of those were from a single county.

The threat to Greater Prairie-Chickens was recognized early on, and the first laws to protect the Heath Hen were passed in 1791. While this effort was too late for the Heath Hen, other early laws (e.g., from 1877) also helped in the management of Greater Prairie-Chickens.

Greater Prairie-Chickens are still hunted in some areas, such as in North and South Dakota, Kansas, Minnesota, Colorado, and Nebraska. In 1999, for example, it is estimated that as much as 30 percent of the total prairie-chicken population in South Dakota was killed during the hunting season.

There is also evidence that the increase in populations of introduced pheasants is harmful to Greater Prairie-Chickens, primarily because of the pheasants laying their eggs in the prairie-chicken nests.

If you are a landowner in the traditional range of the Greater Prairie-Chicken, one of the most beneficial ways you can help the species is to manage your land for it by reducing grazing pressure, adjusting the amount of land in cultivated crops, and controlling burning and tree removal.

The Conservation Reserve Program (CRP) of the US Department of Agriculture (USDA) Farm Service Agency is a voluntary program for agricultural landowners in which they can receive annual rental payments and cost-share assistance to establish long-term, resource-conserving covers on eligible farmland (http://www.fsa.usda.gov/FSA/webapp?area=home&subject=cop r&topic=crp). This program protects millions of acres of topsoil from erosion by promoting the planting of resource-conserving vegetative covers in these areas. This activity restores grasses, trees, and wetlands on millions of acres, which in turn helps the wildlife populations. In July 2013 the USDA announced that, since 2009, it had enrolled nearly 12 million acres in new CRP contracts, and as of July 2013 there were more than 26.9 million acres enrolled through 700,000 contracts with landowners. It is important that federal funding continues to be available for this program so that the number of CRP acres not be reduced. While Greater Prairie-Chickens do poorly if all of the traditional habitat is in cropland, they do very well when about 20 to 30 percent of the land is cropland and the remainder is grassland. In early 2014, the federal farm bill expanded opportunities for conserving grasslands; however, the Agricultural Act of 2014 later decreased the cap on the number of acres that could be enrolled in CRP contracts.

Although most of us are not in the position of being able to manage our land for Greater Prairie-Chickens, we still can help in their conservation and reintroduction. The Adopt-a-Prairie-Chicken program of the Texas Parks and Wildlife Department provides donations to zoos that are raising Attwater's Prairie-Chickens in the hope of restoring wild populations (http://www.tpwd .state.tx.us/huntwild/wild/birding/apc/involvement/).

Another place that your dollars can help is the Conservation Fund (http:// www.conservationfund.org/), which is working to preserve Greater Prairie-Chicken habitat. For example, in Wisconsin, the fund has purchased land next to Buena Vista Wildlife Area to donate to the Wisconsin Department of Natural Resources to manage for prairie-chickens.

It is estimated that in Missouri the number of Greater Prairie-Chickens decreased from hundreds of thousands of birds before European settlement to as few as five hundred birds. The Missouri Prairie Foundation (founded in 1966) owns sixteen tracts of prairie totaling twenty-six acres, which the group actively manages for all prairie species. The organization is also engaged in advocacy for grassland wildlife–friendly agriculture and energy policies, to benefit prairie-chickens and other grassland species, and has an active outreach and education program to inform and encourage Missourians to learn more about prairie. The Missouri chapter of the Nature Conservancy owns and manages several prairie tracts in that state, and one—Wah'Kon-Tah Prairie—is the site of a five-year prairie-chicken translocation project involving the Missouri Department of Conservation (see *Missouri Prairie Journal*, fall–winter 2012, pp. 12–17, available at www.moprairie.org).

A Personal Note: I grew up with Greater Prairie-Chickens. When I was five years old, my parents moved to rural central Wisconsin just north of a Greater Prairie-Chicken sanctuary (Buena Vista Wildlife Area). One year, in fact, a small contingent of Greater Prairie-Chickens actually gathered for their breeding display on a run-down field on our property. I fell in love with them at the sanctuary itself in the springtime, when my high school biology teacher took us there in the predawn hours to sit in a blind and hear the males do their eerie booming long before we could see them. In the winter one could drive the roads of Buena Vista Wildlife Area and see flocks of Greater Prairie-Chickens fly low across the snowy fields or over the road, usually having been disturbed by one of the many wintering Rough-legged or Red-tailed Hawks that gathered at the sanctuary to dine on chicken. In 2006 and 2007 the declining numbers of Greater Prairie-Chickens at Buena Vista were supplemented with forty male and twenty-four female adult Greater Prairie-Chickens, respectively, from northwestern Minnesota.

The Greater Prairie-Chickens in Texas (the Attwater's subspecies) are hanging on by a thread, being continuously and mostly unsuccessfully reintroduced from captive populations raised from eggs gathered from nesting birds in the wild. They do, however, live long enough in the wild to be seen by birdwatchers on spring days early in the morning in the coastal areas in Texas where they have been released. It is to be noted, however, that for a birder who lists species seen, the Attwater's Prairie-Chicken is not considered to be "countable" in Texas because it is no longer a viable, established-in-the-wild subspecies.

### Lesser Prairie-Chicken *(Tympanuchus pallidicinctus)*

Global Population Estimate: 32,000 in 2011, all in North America; an extensive survey in 2013 counted only 17,616

The Lesser Prairie-Chicken is nearly identical to, but somewhat smaller than, the Greater Prairie-Chicken, again being a round, chunky bird with uniformly barred brown feathers and a short, rounded tail. The two species also have a similar display, with the male's inflated neck pouches being slightly

redder in the Lesser Prairie-Chicken and the booming sound being what has been described as a "bubbling, hooting wamp wamp wodum wodum" (as compared to the Greater Prairie-Chicken's "long, low, hooting moan oooa-hooooooooom").

Unlike the preferred habitat of Greater Prairie-Chickens, Lesser Prairie-Chickens are generally found in sand sagebrush–bluestem and shinnery oak–bluestem vegetation areas.

It is estimated that there were about sixty thousand Lesser Prairie-Chickens in the early 1970s and ten thousand to twenty-five thousand in 1999. This is a decline of about 97 percent from the estimated population in the early 1800s. Because Lesser and Greater Prairie-Chickens are so similar and once overlapped in range (that is no longer the case), however, there is some uncertainty as to how many of each there once were. In the nineteenth century, there was no limit on the number of Lesser Prairie-Chickens that could be killed by hunters. In addition, conversion of the prairies to farms and the resultant decrease in native grasslands, as well as serious droughts, fire suppression, oil contamination in Texas, and the spread of woody species into traditional prairie-chicken habitat, have caused the dramatic decrease in the number of Lesser Prairie-Chickens.

As with other prairie grouse species, this decline was recognized many years ago, with the first laws regulating hunting enacted in Kansas in 1861 and shortly thereafter in other states with Lesser Prairie-Chickens. Periodically, these states have banned hunting when the populations were particularly threatened. While their decline is clear and dramatic, some states still allow hunting of Lesser Prairie-Chickens. Limited hunting is still legal in Kansas and Texas. A publication issued by the Texas Parks and Wildlife Department in 2007 estimated a likely extinction of Lesser Prairie-Chickens by 2027 (or possibly as early as 2017).

Southeastern New Mexico near Roswell contains one of the Lesser Prairie-Chicken's most important undisturbed habitats, and it has been designated by the Bureau of Land Management as an area of critical environmental concern. Much of this land has been acquired or otherwise protected for these birds. For example, the Conservation Fund (http://www.conservationfund .org) purchased Sand Ranch, and additional purchases were made with funds from the Land and Water Conservation Fund (www.lwcfcoalition.org) and private foundations.

The Lesser Prairie-Chicken Interstate Working Group completed a plan for the five states (Colorado, Kansas, New Mexico, Oklahoma, and Texas) that still had Lesser Prairie-Chickens in 2008, including guidelines for managing habitat and monitoring the populations of the species. In 2012 the first range-wide aerial survey was done by the Great Plains Landscape Conservation Cooperative together with various federal agencies to assess the numbers of Lesser Prairie-Chickens. The result was a total of 37,170 birds. A second, expanded survey was done in 2013 and found only 17,616 birds, a dramatic decrease, and the number of leks observed had also decreased substantially. Information from this survey will be used as a baseline by the states to monitor trends in the populations and to target conservation programs in partnership with landowners, oil and gas industries, and wind energy and electric utilities.

The High Plains Partnership for Species at Risk was formed to create voluntary grassland restoration projects on private lands. There are US Conservation Reserve and Landowner Incentive programs that are used to help bring landowners into the programs to conserve or place land in suitable prairie habitat. As with the Greater Prairie-Chicken, if you own or use grassland in an appropriate western geographic area, you can manage it for Lesser Prairie-Chickens. Because this species occurs across such a huge territory and information on its historical range and density is incomplete, it is valuable to have reports of sightings of the Lesser Prairie-Chicken. Of course, it is also very important not to harass or otherwise disturb these birds at any time, particularly during courting and nesting. As with other prairie grouse, management of the habitat to enhance grassland cover and limited prescribed burning can be beneficial.

In March 2014 the FWS announced the listing of the Lesser Prairie-Chicken as "threatened," after which the final rule was to be published and then become effective. Various state attorneys general together with gas and oil associations filed suit, claiming that this reclassification was harmful to oil and gas producers, agricultural operations, and the states by increasing the burden of environmental mitigation. Environmental organizations also sued on the grounds that the Lesser Prairie-Chicken should be listed as "endangered" and not as "threatened."

Although there is no question that Lesser Prairie-Chickens are in trouble, there are numerous lobbying pressures and other obstacles standing in the way of listing the Lesser Prairie-Chicken as endangered. You can help in the effort to get the plight of the Lesser Prairie-Chickens recognized by contacting your legislators and asking them to have the Lesser Prairie-Chicken listed as "endangered" under the Endangered Species Act rules.

The Playa Lakes Joint Venture (PLJV, at http://www.pljv.org) is a non-profit partnership of federal and state wildlife agencies, conservation groups, private industry, and landowners dedicated to conserving bird habitat in short-grass and mixed-grass regions of the Southern Great Plains (eastern Colorado and New Mexico, western Nebraska, Kansas, Oklahoma, and the Texas Panhandle). This group provides science-based guidance, decision support tools, outreach, coordination, and financial support. One of the birds for which it aims to conserve habitat is the Lesser Prairie-Chicken, a bird found nowhere else but in this group's target geographic area. The PLJV invites anyone who is interested to become involved in a local conservation partnership. A Personal Note: My move to Texas in 2000 gave me a chance to see my first-ever Lesser Prairie-Chickens in the Texas Panhandle. I was lucky enough to be able to see them for the ten years that I lived in Texas by always scheduling a Panhandle trip in early spring. When I found that the chamber of commerce in Canadian, Texas, sponsored early-morning trips to blinds strategically placed on ranchers' lands, I signed up. Along with a small group of other birders, I was trucked out across uneven ranch roads to a blind, where we sat on hay bales and waited in hiding until the male Lesser Prairie-Chickens appeared. The males sought out an open, sometimes raised area to begin their displays. Periodically, one of them would come very close to the blind and sometimes even hop up on the blind above our heads. As the sun started illuminating the area, there was only a short window of time when it was light enough for good photographs before the birds quit displaying for the day. A couple of the years I found the Lesser Prairie Chickens on my own by driving the roads north of Canadian on very early mornings in the spring, then listening and looking for their round bodies out in the grassy fields. One year, I almost witnessed the demise of one of these rare birds as I watched two Rough-legged Hawks hunting over a field where I had just seen a couple of Lesser Prairie-Chickens, but that day at least the chickens won.

## Yellow-billed Loon *(Gavia adamsii)*

Global Population Estimate: 23,500, of which about 16,000 are present on the North American continent (the Natural Resources Defense Council [NRDC] estimates 3,000 to 4,000 in Alaska of a world population of 17,000).

The Yellow-billed Loon is a large, nearly goose-sized water bird. Like the similar but smaller, more numerous, and well-known Common Loon, the Yellow-billed Loon in breeding plumage has a blackish-green head, white breast, and a back with a rectangular white-and-dark pattern. As indicated by the species name, the large, upturned bill of the Yellow-billed Loon is a straw-yellow color, unlike that of the Common Loon. In winter the Yellow-billed Loon loses the dark coloration on the head and becomes a buffy brown-and-white bird overall, but it still has a yellowish bill.

Yellow-billed Loons are found on lakes and slow-moving rivers, usually near the coast. They nest on lakeshores and islands in the remote high Arctic, usually on tundra in far northern Canada east to Hudson Bay, and in Siberia and Alaska, with about one-quarter of the global population nesting on lakes on the North Slope of Alaska. Preferred lakes have low shorelines and stable water levels. Yellow-billed Loons are very territorial and defend their portion of a lake, or the whole lake, from intruding loons and diving ducks.

When nesting is finished for the season, Yellow-billed Loons migrate in pairs and small flocks to warmer coastal waters, such as areas off of southeastern Alaska, British Columbia, and Washington State, as well in the coastal waters of Japan, China, and the Korean peninsula. They can periodically be

found far inland, particularly on large reservoirs. Their diet is primarily fish, which they dive to get (loons are called "divers" in Europe). Before lake ice has melted in the spring, Yellow-billed Loons dive beneath the ice to find food. They also eat mollusks, crustaceans, and other aquatic animals.

Because most early naturalists rarely wandered as far north as Yellow-billed Loons are typically found, it is unlikely that there are any accurate estimates of the earlier Yellow-billed Loon populations, at least in North America.

Yellow-billed Loons have apparently never been numerous; however, breeding ground surveys indicate that the population is decreasing, for example, by about 5 percent between 2000 and 2004. The size of the Yellow-billed Loon population is small enough that habitat loss or disruption, such as by oil exploration or oil pollution on its breeding grounds or in offshore coastal waters, or disturbance by humans, such as drilling activity on the Alaskan North Slope, hunting by natives for food or other reasons, or reduction in the numbers of fish, can threaten the population substantially. Unfortunately, an area of the Colville River delta in Alaska that was identified as an important area where Yellow-billed Loons should be protected from development was later leased for oil exploration and production.

There are habitat protection provisions of a migratory bird treaty with the former Soviet Union that could benefit Yellow-billed Loons if the treaty were implemented.

Because of the remoteness of the areas where Yellow-billed Loons are found, most people cannot directly improve the situation for them. Global actors, such as the US government and energy companies working on North Slope exploration and drilling, need to take the Yellow-billed Loon into account.

In March 2014 a finding by the FWS that Yellow-billed Loons should be listed as endangered was published; however, the FWS stated that the actual listing of the Yellow-billed Loon as endangered would not happen until other, higher-priority listings occur.

Concerned citizens staying aware of legislation and contacting their legislators is probably the best hope for the Yellow-billed Loon. In addition, the Natural Resources Defense Council (http://www.nrdc.org/), which has fought to get protected status for Yellow-billed Loons, can be supported.

A Personal Note: In spite of having birded most of my life, I can almost count the number of Yellow-billed Loons that I have seen on one hand. My first sighting was of a single loon on a pelagic (ocean) trip off the California coast in the 1990s. Although it was a very welcome life bird for me, I only realized how rare it was when Debi Shearwater, the trip organizer, had me (one of the few people on board with a long camera lens) hang off the edge of the boat and take as many pictures as possible to document the sighting. After I moved to Texas in 2000, I almost got to see one that was being spotted in the waters off of South Padre Island, but I delayed too long (that was before I became a very serious "chaser" of bird rarities) and it was gone before I could travel the 525 miles down there from Fort Worth. As is typical of some other rare birds discussed in this book, Yellow-billed Loons only rarely appear in areas where humans find them, and they do not usually stay around long enough to be seen by many people.

My second sighting was in June 2008, when four Yellow-billed Loons were flying low over the water off of St. Lawrence Island, Alaska, heading north. It was my only sighting of them that year, although I spent much time staring out over the waters surrounding Alaska.

My most recent sighting was in spring 2010, when a single Yellow-billed Loon was diving on a lake in Oklahoma City. This bird was not an adult, and perhaps due to its drabness, as well as wind whipping the waves across the lake, it took me two trips from Fort Worth before I could find the bird. Distant Common Loons wintering on the lake needed to be checked to be sure they weren't their somewhat larger, yellow-billed cousins before I could be sure. Finally, after one more drive around the lake, I found the Yellow-billed Loon quite near the shore, where I could photograph it whenever it came up from its dives.

## Clark's Grebe *(Aechmophorus clarkii)*
Global Population Estimate: 15,000, all in North America

Clark's Grebes were once classified as Western Grebes, a species that is much more numerous, but differences in plumage, as well as different vocalizations, caused them to be classified as a separate species in the late twentieth century.

Both Clark's Grebes and Western Grebes are large waterbirds with long white necks and black coloring extending up the back of their necks and terminating in a black cap. They each have long, thin, pointed bills, with the bill of Clark's Grebe being bright yellow and the bill of the Western Grebe being greenish-yellow. The other main difference between these two species is that in breeding plumage, the Clark's Grebe has a white face with white around

the eye, while the lower portion of the Western Grebe's black cap surrounds the eye. The winter face pattern, however, is quite similar in the two species. When the birds are swimming in the water, the flanks of the Clark's Grebe are generally pale gray, while the flanks of the Western Grebe are darker gray. Both grebe species are well known for their unique dual courtship displays in which a side-by-side pair of birds seems to run across the top of the water with their long necks curved stylistically and their beaks pointed up at the same angle.

Very little is known about the historic populations of each species because they were so recently classified together and are so similar. They both spend most of their time in the water of open bays near marshes, where they nest. The best nesting sites have large clumps of water plants such as bulrushes that emerge from the water and are surrounded by water. Neither species flies very much except during migration, usually only moving to another area of the water when disturbed. Because they do not leave their water habitat to feed, there must be a sufficient number of fish in the water to sustain the grebe population.

From shortly after Audubon's death in 1851 until the early 1900s, both Clark's Grebe and the very similar Western Grebe were shot for their white undertail plumage, which was used for capes, coats, and hats. In some cases hunting eliminated these species from particular areas. From the late 1800s until the early 1900s tens of thousands of the two grebe species were shot.

While it is no longer legal to hunt Clark's Grebes, the grebes still face major problems due to oil spills and gill nets. In addition, floating debris, such as nylon fishing lines, fishing lures, and other plastic and rubber products, can entangle the birds and pollute their territories. When Clark's Grebes are disturbed, whether by boaters or other human activities, they leave their nesting area, thus abandoning nests or failing to nest altogether.

It is also important that water levels in the Clark's Grebe habitats be sufficiently high to provide open water so that the nesting habitat is not easily accessible to predators but not so high that the emergent vegetation and nesting sites are covered. It is important that lake habitats for these grebes not be drained, allowed to go dry, be developed, or be subject to human disturbance.

Audubon California (http://ca.audubon.org) has a project to protect breeding Clark's and Western Grebes at four California lakes, and it is funded by settlements of lawsuits related to oil spills that harmed grebes in the ocean. This project involves local California Audubon chapters working to monitor the grebes, reduce human disturbance on the lakes, keep the water levels appropriate for grebe breeding, and determine what other factors might be detrimental to the grebes.

A Personal Note: Both Western and Clark's Grebes nest in Texas, as I was delighted to learn when I first ventured west from Fort Worth to Lake Balmorhea near Fort Davis, some eight hours' drive away. Because this is a large lake and a telescope was often needed to see all the grebes, each time I visited the lake it would take me a long time to be sure that, among all the Western Grebes, I was also seeing a few Clark's Grebes.

---

BLACK-CAPPED PETREL AND BERMUDA PETREL

Black-capped and Bermuda Petrels are seabirds with dark and white coloring. They silently fly over the waves with stiff wing-beats and long, arcing glides.

## Black-capped Petrel *(Pterodroma hasitata)*

Global Population Estimate: 5,000, in the waters of the southeastern United States and northern South America

The Black-capped Petrel is about sixteen inches long and belongs to the "gadfly petrels," a group of small seabirds that typically feed by picking food off the surface of the seawater. Like many seabirds, Black-capped Petrels are basically dark brownish-black in color with a white belly, but their dark cap, white collar, white rump, and white underwings with a diagonal dark stripe can be used to separate them from other seabirds, even by an observer on a tossing, rolling ship. They also have a wedge-shaped tail and long, pointed wings, which they usually hold in a bent sickle shape. They fly erratically, with a rolling, agile arcing flight, especially in heavy winds.

Beginning in early November, Black-capped Petrels breed on Caribbean islands, where this species is known as the *diablotín* (little devil) because of their devilish, screaming calls. For nesting, they excavate burrows in soil or use cracks in rocky areas. Because of their low numbers and because they are usually far out at sea, they are rarely seen even on most pelagic trips, though they are regularly seen on North Carolina pelagic trips in the summer. In the nonbreeding season, Black-capped Petrels live at sea, feeding in deep, nutrient-rich areas where they can find sargassum, such as at the western edge of the Gulf Stream and in Gulf Stream eddies.

Black-capped Petrels once were common in their breeding range in the Caribbean. Monitoring the populations of Black-capped Petrels has always been difficult, however, because they nest on remote island areas and only come to their nesting burrows at night. There is strong evidence of a dramatic decline in numbers, however. From 1961 to 1987 the number of breeding pairs is estimated to have declined about 40 percent.

It was announced in June 2014 that transmitters had been placed on Black-capped Petrels captured from their burrows in remote areas of the Dominican Republic in order to gather data on the species. This effort, supported by the US Geological Survey, Clemson University, Grupo Jaragua in the Dominican Republic, and American Bird Conservancy, allows researchers to track the daily movements of this little-known species.

One can become better informed about the species by taking a summer pelagic trip to the Gulf Stream off of North Carolina's Outer Banks and can support organizations that recognize the precarious status of Black-capped Petrels. These include BirdLife International (http://www.birdlife.org/) and the American Bird Conservancy (http://www.abcbirds.org/). The Seabirds program of the American Bird Conservancy specifically aims to solve seabird problems by means such as educating persons engaged in commercial or recreational fishing on better techniques to reduce seabird deaths, eradication of non-native harmful animals on islands, and removal of invasive island vegetation.

A Personal Note: For people who have not gone out on a boat seeking pelagic bird species, it is difficult to imagine how hard it is to confidently discern the difference between distant, small, black-and-white blurs, with the boat rocking and tipping, water washing across the boat and dripping from the boat overhang, eyeglasses and binoculars becoming crusted with salt, and the binoculars refusing to keep the bird in view and in focus.

When I lived in North Carolina, I took my first pelagic trip, and many more after that. Because we usually saw Black-capped Petrels on these North Carolina trips, I did not appreciate their rarity until I began to go on pelagic trips elsewhere.

## *Bermuda Petrel* (*Pterodroma cahow*)

Global Population Estimate: 180 in 2011; "over 100 nesting pairs" in 2012, primarily near Bermuda

The Bermuda Petrel, the national bird of Bermuda, is one of the rarest seabirds in the world. Like the Black-capped Petrel, the Bermuda Petrel is a nocturnal bird that traditionally nests in burrows. In Bermuda, this species is called the cahow, a word likely derived from the bird's eerie cries. Bermuda Petrels are slightly smaller than Black-capped Petrels and have a similar appearance with a cap, but instead of a white collar they have a brown collar, and they have a much smaller white area on the rump.

The only place that Bermuda Petrels breed is on a few small, rocky islands in a single harbor in the Bermuda island grouping. Bermuda Petrels, like many other seabirds, spend all their time at sea before they nest at about five years of age, and after that, when they are not sitting on their nests, they remain at sea. They eat primarily small squid, shrimp, and fish.

Bermuda Petrels almost became extinct in the 1600s because early colonists hunted and ate them, destroyed the sites where they bred, and introduced harmful predators (rats, dogs, and cats). In fact, until they were rediscovered in 1951 they were believed to be extinct.

The Bermuda Conservation Programme removed the rats and installed petrel-sized burrows made of concrete with the openings restricted to keep out White-tailed Tropicbirds. In addition, a nearby island was restored as a place for Bermuda Petrels to nest.

Because there are so few Bermuda Petrels, any kind of disturbance

—whether it be reintroduction of predators, sea level rise, or hurricanes—can still threaten their entire population. In addition, as with many other seabirds, Bermuda Petrels become ensnared in or eat plastic and other trash that is washed out to sea from beaches. They are also susceptible to becoming oiled and flightless when they sit on water covered with oil that has washed into the sea from city sewers or has leaked from containers or pipelines.

Although Bermuda Petrels once bred in burrows in soft, sandy areas, introduced predators have made such habitat unsafe and off limits. The creation of artificial burrows and areas with eroded limestone provide the only places where Bermuda Petrels now nest. In the early twenty-first century, various groups, including the National Audubon Society Seabird Restoration Program (http://conservation.audubon.org/seabird-restoration-program), have worked to increase the number of artificial burrows and to use the sound of courtship cries to attract Bermuda Petrels from their destroyed nesting sites (rendered useless by recent hurricanes) and, when the birds are returning from the sea, to the new sites. Chicks were moved to the new sites and fed there with the goal of getting them imprinted with the new location and thus returning there to nest.

In addition to cutting up plastic waste and properly discarding it and refraining from polluting waters, proactive action can include helping clean up Bermuda beaches. To learn more, and possibly become directly involved in helping Bermuda Petrels, one can join the Bermuda Audubon Society (http://www.audubon.bm).

Although most likely premature in its conclusion, an American Bird Conservancy article entitled "Bermuda Petrel Saved from Extinction," released on April 12, 2012, reported that there are "more than 100 nesting pairs" of Bermuda Petrels, which indicates that the Cahow Recovery Program managed by the Bermuda Department of Conservation Services and the other programs mentioned above are making a difference.

A Personal Note: In 1999 I took one last North Carolina pelagic trip before I moved to Texas, and I was expecting to see only the usual pelagic species. On this trip, however, I definitely realized how rare Bermuda Petrels are when Brian Patteson, our boat captain and chief bird guide, nearly fell off the boat in his excitement when a Bermuda Petrel was spotted. So far, that is the only one I have seen, and that may remain the case.

## Ashy Storm-Petrel *(Oceanodroma homochroa)*

Global Population Estimate: 5,000 to 15,000, only on the islands off California and adjacent waters

Storm-petrels are small seabirds (usually five to seven inches long) that feed by seeming to walk or run on the water as they pluck small sea creatures such as small fish, squid, and crustaceans off the surface. The Ashy-Storm Petrel is an all-over dark gray-brown ("ashy") bird with a forked tail. It mainly feeds at night and has a flight style that is called "fluttery." It breeds in about seventeen different locales hidden on the talus slopes of rocky islands off the coasts of California and northern Mexico. It only infrequently travels far from where it breeds and is unlikely to be seen from the mainland. Of the estimated global population of Ashy Storm-Petrels, probably half of them come to Monterey Bay and the surrounding area in the fall, where huge sardine and anchovy populations are their food sources.

It is thought that the population of Ashy Storm-Petrels was stable and the species was not endangered from at least the 1880s until the 1970s, based on records from the Farallon Islands. Since then, however, the data indicate that the population size has declined precipitously, apparently primarily due

to pollution and introduced and natural predators on their previous nesting islands. Further decline is likely to be due to sea level rises and changes in upwelling. In addition, introduced grasses have overgrown many of the nesting areas, making them unusable by the Ashy Storm-Petrels. While the estimated world population is smaller than a number of other species that are federally listed as threatened, the Ashy Storm-Petrels in the United States are relatively safe because they all nest on islands that are protected from disturbance and development by humans. Although Ashy Storm-Petrels were under consideration for protection under the Endangered Species Act, at the end of October 2013 the FWS denied such protection for a second time.

In addition to plucking their natural food from the sea surface, Ashy Storm-Petrels also pluck bits of plastic and other debris discarded by humans. Ingestion of such indigestible debris means that the petrels are unable to eat sufficient amounts of their usual food, which threatens their health. The small localized population of Ashy Storm-Petrels makes the population more susceptible to problems caused by bad weather; island predators such as Western Gulls (which have increased dramatically on some of the Ashy Storm-Petrels' nesting islands), rodents, and cats; and pollution. Nest monitoring by researchers has caused problems too, as the birds are easily disturbed.

Ashy Storm-Petrels will use artificial nest boxes, which provide them with additional nesting sites and which can be placed to cover and protect nest chambers that have new fledglings.

Because the US national parks are important in the protection of Ashy Storm-Petrels, it is important that citizens actively support funding for the park service. Island Conservation, a nonprofit group, works to support habitat for seabirds, including the Ashy Storm-Petrel, on islands off of California and Baja California, Mexico (http://www.islandconservation.org/). The American Bird Conservancy also has an Ocean and Islands Seabirds Group (http://www.abcbirds.org/abcprograms/oceansandislands/seabirds.html). A Personal Note: I saw Ashy Storm-Petrels on only one Californian pelagic trip, but I had no doubt (unlike other occasions) that I was actually "on" the correct birds that the experts were calling out as the boat swayed and pitched on the waves. It did not hurt that there were many Ashy Storm-Petrels. The flocks of little dark birds dotting the ocean surface off of Monterey, California, also included other storm-petrels, but with a bit of help I was able to figure out which ones were the Ashy Storm-Petrels by their color and flight style.

## Reddish Egret (*Egretta rufescens*)

Global Population Estimate: 67,500, in the southern United States and Latin America

Reddish Egrets stand about two and a half feet high and have a four-foot wingspan. They are somewhat smaller than the similarly shaped Great Egrets and substantially smaller than Great Blue Herons. Although many Reddish Egrets do actually have an elongated, reddish-cinnamon head and neck with a dark, slate-blue back, other Reddish Egrets are not reddish at all but as white as other egrets. The dark and the white color morph (form) can be found in the same family. The white Reddish Egrets can be distinguished from other white egrets by their size and by their pink-based dark bill (in breeding plumage) and slate-gray legs.

The distinctive feeding behavior of Reddish Egrets also is helpful in distinguishing the white ones from other egrets. Reddish Egrets feed mainly on small fish and are known for their unique feeding behavior, in which they spread their wings as they dart erratically over shallow water after fish. They also are known to elevate their wings into sort of an umbrella shape to create a shady area that attracts fish while the egret waits motionlessly. Reddish

Egrets also periodically stir or scrape the mud with a foot and then peer at the water to look for prey. They also regularly take advantage of pools that appear when the water level is going down during drought and the fish are trapped and easier for the egrets to catch.

Reddish Egrets are found only in southern coastal wetland areas, such as mangrove keys, shallow lagoons, and islands with low vegetation. The US populations are thus located primarily along the Gulf of Mexico. Reddish Egrets are also found on both coasts of Mexico and on Caribbean islands. The dark form is more common in the US populations, with the white form being more common in the Bahamas and Greater Antilles. Although Reddish Egrets do not regularly do any type of long-distance migration, they are known to wander somewhat from their breeding area after they have finished nesting.

In the 1800s and into the early 1900s, Reddish Egrets were shot for their plumes. Nestlings and eggs were also taken, so Reddish Egrets were almost eliminated completely from the country. Development in many of the coastal areas has also reduced the habitat that can be used by Reddish Egrets. Although they may nest in wetland areas near human habitation, they may abandon nests if there is too much disturbance.

Reddish Egrets are now the rarest egret in the United States. The national population of Reddish Egrets is believed to be about 2,000 pairs, 1,500 of which are in Texas, 350 to 400 in Florida, 50 to 150 in Louisiana, and a few in Alabama. There are only small numbers of Reddish Egrets in Mexico, the West Indies, and the Bahamas.

A Personal Note: Although I was lucky enough to see a few of the white morph Reddish Egrets that came to the North Carolina coast when I lived there, it wasn't until I moved to Texas that I was able to see the more rusty-hued Reddish Egrets. Usually when I went to the lower or central Texas coast I looked for and often found one or two long-legged, gangly reddish and gray Reddish Egrets patrolling the shallow wetlands, sometimes in big puddles along the road, rapidly moving back and forth with their wings spread in unpredictable, comical dashes across the water.

## Ferruginous Hawk *(Buteo regalis)*
Global Population Estimate: accurate numbers do not appear to be available

The Ferruginous Hawk is a large, long-winged hawk with distinctive, rusty-colored feathered legs and back (the word *ferruginous* comes from the Latin *ferrugo* or *ferrum*, meaning iron rust) and a large white area on the flight feathers. Although it is often mistaken for the common Red-tailed Hawk, the Ferruginous Hawk is so large that it can also possibly be mistaken for an eagle. The head of a Ferruginous Hawk is typically whiter than that of a Red-tailed Hawk. One distinctive feature of a Ferruginous Hawk is the extended yellow gape behind its beak (assuming that you can get close enough to see this feature). Ferruginous Hawks are most likely to be found in western prairies and grasslands, as well as plateaus and shallow canyons, from northern Mexico to southwestern Canada, where they dine on small mammals (rabbits) and reptiles, amphibians, and insects. They sometimes hover, or they may dive from a perch or hunt on the ground.

Ferruginous Hawks were once widespread over western grasslands, but it is unknown how common they were. In the early twentieth century, eggs of Ferruginous Hawks were collected, and the birds were also hunted.

The spread of agriculture and trees on the prairies and the increase in towns and subdivisions have encroached on the territories of Ferruginous Hawks, and in some cases the hawks have been extirpated from their traditional areas. The destruction of many prairie-dog towns has eliminated one of their major foods in some areas and has caused further decreases in the numbers of Ferruginous Hawks. For example, in Canada it was estimated that by 1980 the range size had decreased by 48 percent. Ferruginous Hawks are considered threatened in several states. There has been an increase in their numbers in some areas, particularly where landowners have realized the value of having a rodent predator flying over their fields. In other areas, the population has declined due to cultivation of grasslands and the subsequent reduction in rodent habitat.

There is at least one preserve where Ferruginous Hawks are found. At the Boardman Grasslands in Oregon these hawks share habitat with (and probably eat) the endangered Washington ground squirrel (http://www.nature.org/ourinitiatives/regions/northamerica/unitedstates/oregon/placesweprotect/boardman-grasslands.xml).

While I have not located any specific organizations devoted to Ferruginous Hawks, participating in National Audubon's Christmas Bird Counts (http://www.audubon.org/bird/cbc) can help to document the status of the species. In states where Ferruginous Hawks are found, the designation of Important Bird Areas (http://www.audubon.org/bird/iba) helps to manage and conserve appropriate habitat.

A Personal Note: In winter it is relatively easy to find a Ferruginous Hawk if you go to the Texas Panhandle. Some winters they can even be found in north-central Texas, but they can easily be overlooked amid all the wintering Red-tailed Hawks. It thus took me a couple of years after I moved to Texas to find and identify my first Texas Ferruginous Hawk. A few hints from the bird pros, however, helped me to realize that the large, light-tailed "red-tails" with large whitish patches on the upper wings were Ferruginous Hawks. It was easier to see what they were when they perched on utility poles, because then their rusty backs and leggings were more visible. After I moved to South Dakota, I was delighted to see young, fluffy Ferruginous Hawks on their nest, which was on the ground, because their nesting area had no trees. In other areas of the country, the nests of Ferruginous Hawks are found in trees and shrubs.

YELLOW RAIL AND BLACK RAIL

Yellow Rails and Black Rails are tiny, secretive, nearly impossible-to-see species that are found in damp marshes and grasslands.

## Yellow Rail *(Coturnicops noveboracensis)*

Global Population Estimate: 17,500, all in the United States and Canada

Yellow Rails are the second smallest North American rail (just over seven inches long), with the smallest rail being the Black Rail. Both Yellow Rails and Black Rails live in dense marsh vegetation and are rarely seen or flushed at all, even if one walks through the marsh. Many experienced birders have never seen either one of these rails.

Yellow Rails are small, buffy, mousy birds with a short, pale bill. They are usually found in wet areas, including boggy areas, wet meadows, and other freshwater or brackish areas, preferably with sedges or grasses. They mainly eat snails, aquatic vertebrates, and seeds. Although Yellow Rails do not regularly vocalize, during breeding season the males do make a five-note clicking call that sounds like pebbles tapping together: "*click-click, click-click, click.*"

Yellow Rails nest in Canada and the northern United States, primarily east of the Rocky Mountains. The places where they breed usually have shallow standing water in the spring but no water or very shallow water by the end of their nesting season. Yellow Rails generally migrate south to winter in drier areas in stands of spartina (cordgrass) in coastal marshes.

Because of their small size and their scattered populations, and because they are usually silent and stay hidden in the sedges and grasses and do not fly even when people walk through their habitats, it is extremely difficult to know how many Yellow Rails there were, or are.

While there do not appear to be organizations or efforts devoted specifically to Yellow Rails, support for the national and state wildlife refuges where Yellow Rails are found helps limit wetland development in those areas and restricts access during their breeding season. An example of a prime wintering habitat for Yellow Rails is Anahuac National Wildlife Refuge in Texas.

In states where Yellow Rails are found, you can support conservation easements maintained by nonprofit groups, individuals, and the government, as well as other habitat management for preservation of appropriate wetlands for the rails. Ducks Unlimited, discussed below with respect to Black Rails, works not only to provide wetlands suitable for ducks but also to conserve a wide variety of wetlands, some of which are also suitable for Yellow Rails. A Personal Note: Although I cannot personally verify this fact, Yellow Rails breed in Wisconsin, which is where I grew up and began my birding life. It wasn't until I moved to Texas, where many Yellow Rails winter, that I had a real chance to see one, however. Each spring, David Sarkozi (former president of the Texas Ornithological Society) volunteers to conduct "Yellow Rail walks" at Anahuac National Wildlife Refuge, east of Houston. In order to get the skulking, wintering little rails (both Black and Yellow) to show themselves above the clumps of wet marsh grass and sedges, a rope weighted at intervals with rock-filled plastic bottles is bouncingly dragged between two people across a likely section of marsh. Eager birders wearing tall rubber boots do their best to keep up with the quickly advancing rope without tripping over the sturdy vegetation clumps and falling into the standing water between them. The lore is that someone must fall (often me), as an offering to the gods, before a Yellow Rail will be seen (since someone always falls early in the walk, this theory has yet to be disproved). Each of the birders wears binoculars of course, and some of them also carry a camera and telephoto lens in the persistent hope that they can snap a picture as a tiny, buffy bird with white wing-patches finally lifts off above the vegetation and coasts, usually for only a few feet, before dropping down and disappearing again. I saw this happen most of the eleven springs that I lived in Texas, a few times twice in the same spring, but I have never gotten a picture of a Yellow Rail. A much better way to see Yellow Rails (I am told) is to become friends with a rice grower in coastal Texas and ride along on the rice-harvesting equipment. Even then, the rails apparently fly only when all the vegetation cover is removed, and when they do they don't go far.

## Black Rail *(Laterallus jamaicensis)*
Global Population Estimate: unknown

The Black Rail is the smallest North American rail (about six inches long and weighing 1.1 ounces) and is at least as secretive as the Yellow Rail in its actions. As a Black Rail runs through marshes like a mouse or sparrow, it is usually not detected. Black Rails appear to be more vocal than Yellow Rails, however, with their "*kick-ee-doo*" or "*kee-kee-krr*" three-note call heard mainly at night. As its name indicates, the Black Rail is basically black with a short black bill. Black Rails eat insects, crustaceans, and other little marsh animals.

Black Rails breed much farther south than the Yellow Rail, primarily in coastal salt- or freshwater marsh areas (for example, coastal California and along the Gulf of Mexico) where the ground is wet but not completely underwater. There are also widespread but local inland breeding areas from Colorado, Kansas, Oklahoma, and the upper Midwest (Minnesota, Illinois, Michigan) to the East Coast.

Black Rails migrate short distances at night in the spring and fall. Generally, Black Rails winter in the southern areas of their breeding range, primarily in central Florida, the Gulf Coast from Texas to Florida, and southern California.

Black Rail populations have been decreasing for more than a century, possibly by as much as 75 percent since the 1990s, primarily due to habitat loss. They may breed only in a small number of places in each coastal state where they once were much more numerous, and they have disappeared as well from a number of interior sites where they formerly bred. Individual populations of Black Rails have disappeared (e.g., in Massachusetts and some places in California) or declined dramatically. Population size is very difficult to estimate due to the difficulty of seeing Black Rails.

As with Yellow Rails, Black Rails are in trouble due to loss of wetland habitats, particularly salt marshes. Such marshes are regularly dredged or drained for mosquito control. Lining irrigation canals eliminates the shallow wetlands that are preferred by Black Rails. The introduced phragmites reed has also invaded many marshes, making the habit less suitable for rails. Because many Black Rail habitats are near the ocean, the rise of sea levels threatens them as well, and they are already seriously in danger, particularly along the East Coast and in San Francisco Bay, where most of the tidal marshes have been destroyed due to agriculture, salt production, and urban development. It is expected that without substantial efforts to preserve and improve wetland habitats, the number of Black Rails will further decrease as their habitat shrinks.

There are local groups along the coasts, such as California's Marin County Audubon Society (http://www.marinaudubon.org/), that are engaged in efforts to restore and preserve wetlands, including those needed by Black Rails. The Center for Conservation Biology (in Williamsburg, Virginia) has an Eastern Black Rail Conservation and Management Working Group (http://ccb-wm.org/BlackRail/index.htm) that is doing a status assessment and a conservation action plan for the Black Rail. Assessing the status of the

Black Rail is particularly important because there currently are insufficient data to indicate with any certainty whether the number of Black Rails is decreasing and if so, how much.

An organization that has done and continues to do much to conserve wetlands is Ducks Unlimited, which has as its mission the conservation, restoration, and management of wetlands and associated habitats for North American waterfowl. Rails are not classified as "waterfowl," but some of the efforts made by Ducks Unlimited for ducks and geese also benefit rails. Thus, Ducks Unlimited works to restore grasslands and watersheds, acquire land, and establish conservation easements and wetland management agreements so that wetlands remain suitable for wildlife such as rails. See http://www .ducks.org for more information on what Ducks Unlimited is doing and how you can support its efforts.

A Personal Note: I have seen a couple of Black Rails briefly on the Yellow Rail walks discussed above, but my first sighting was back when I lived in North Carolina. A group of birders on a Carolina Bird Club field trip tromped around a couple of coastal area marshes looking for rails in general, including Black Rails. After we had seen a Virginia Rail, a tape was played of the distinctive Black Rail call to see if we could get a response from somewhere in the low, sedgy area where the group was birding. Immediately we could hear the "*ki-ki-krr*" of an invisible interested Black Rail—my first time to hear a Black Rail. The bird's hormones were apparently in overdrive and the bird rushed toward us, completely unseen in the sedges, calling excitedly. All the birders rushed to try to see it, and we surrounded the sound and then gradually moved toward it in a slowly tightening circle. The bird kept calling, but no one could see it. Finally we were nearly boot-to-boot in a small circle, but there was no sighting of a rail, and then no sound. In horror, I realized that one of us had probably crushed the tiny, mouselike bird. Before I had a chance to feel totally guilty, the rail started calling just outside our circle! Then we all saw the minuscule black bird darting across a small patch of mud between sedge clumps. It appeared to be trying to locate the taped rail call that it had heard. Since that time, although I have heard Black Rails in marshy areas of Texas, with some of them surely having been less than a few feet from me, the only time I saw others was when I got microglimpses of skulking Black Rails during Yellow Rail walks at Anahuac National Wildlife Refuge.

### Whooping Crane (*Grus americana*)

Global Population Estimate: wintering population, 2010–11: 437 in the wild (279 in the Aransas–Wood Buffalo flock discussed below; others being introduced into the wild); early 2014: about 593, all in the United States and Canada

Whooping Cranes are tall (about five feet), long-legged, long-necked, and magnificent birds. Adults are white with a red patch on their heads and a white feather bustle. When they fly, their black wing-tips can be seen. Their name comes from their calls, variously described as a shrill, trumpeting sound or "*ker-loo ker-lee-loo*."

While walking in shallow water or in fields, Whooping Cranes feed on crustaceans, fish, berries, and other small aquatic animals and plants. Migrating cranes regularly feed on waste grain such as wheat, barley, and corn.

Whooping Cranes require inland freshwater wetlands for breeding and coastal brackish wetlands for their winter habitat. Although their winter habitat is primarily Aransas National Wildlife Refuge and nearby coastal areas in Texas, six Whooping Cranes wintered in 2011–12 on Granger Lake in Central Texas, possibly due to very warm winter weather. Most wild Whooping Cranes breed in Canada in northeastern Alberta at Wood Buffalo National Park, a huge, poorly drained muskeg area with many shallow-water wetlands.

An estimated 15,000 to 20,000 Whooping Cranes once were found from central Mexico to the Canadian Arctic and from the East Coast west to Utah. Hunting of Whooping Cranes from 1870 to 1920 was documented as the cause of 254 deaths, greater than their estimated annual reproduction at that time. Hunts for other large birds such as Sandhill Cranes and Tundra Swans also resulted in killing of Whooping Cranes. Even more critically, the conversion of much of the Great Plains to agricultural use caused Whooping Cranes to disappear from the Northern Great Plains in the United States and Canada, and construction and use of the Gulf Intracoastal Waterway eliminated about 11 percent of their wintering habitat. By 1941 there was a single small flock that migrated annually, usually in family groups, between the Wood Buffalo breeding grounds in Canada and the wintering grounds at Aransas National Wildlife Refuge in Texas.

Since 1941, when the population reached an all-time low of fifteen to twenty-one birds, extensive efforts have been made to increase the population

LEB

of Whooping Cranes. Captive propagation was used at the Patuxent Environmental Science Center in Patuxent, Maryland, to produce fertile eggs and establish a captive flock in Baraboo, Wisconsin, at the International Crane Foundation (ICF) and at the Calgary Zoo in Alberta. The ICF (http://www.savingcranes.org) works to support research on cranes worldwide and to educate people about cranes and their habitats and status with conservation activities to manage and restore crane habitat. In addition, through their on-site captive crane breeding program, the ICF is working to reintroduce cranes, including the Whooping Crane, into the cranes' traditional breeding and wintering areas.

Efforts have also been made to improve habitat along the Platte River in Nebraska, where many Sandhill Cranes stop over during spring migration, and agricultural crops have been planted at various state and national wildlife refuges.

In 2010, 74 wild pairs of Whooping Cranes nested and fledged 49 chicks, 45 of which arrived at their winter homes at or near Aransas National Wildlife Refuge. As of May 5, 2011, there were 279 Whooping Cranes in the wild, of which 44 were young of the year and 156 were paired adults. By early 2014 the estimate for the total number of Whooping Cranes was almost 600. Also, in spite of extensive publicity on the illegality of harming Whooping Cranes, there are regular incidents in which Whooping Cranes have been found to have been shot and killed. In February 2014, 3 Whooping Cranes were reported as having been shot on their wintering range, bringing to at least 16 the total number of Whooping Cranes that had been shot in a five-year period. There is always a need for more education of people in the migration and wintering ranges of Whooping Cranes.

In addition to the main population of Whooping Cranes, there are three other populations being established: a Florida nonmigratory population (20 birds), a Louisiana nonmigratory population (10 young birds), and a Wisconsin-Florida migratory population (105 birds, of which 88 are adults).

While intensive efforts at conservation and flock protection allowed the numbers of cranes in the Wood Buffalo–Aransas flock to increase to more than two hundred by 2005, having only one self-sustaining flock means that any weather (e.g., a hurricane on the Texas coast) or food or health disaster can wipe out the entire flock at once. There are also concerns about heavy barge and other boat traffic in Texas, airplane flights over crane wintering and breeding grounds, and pollution in the Gulf Intracoastal Waterway, which passes through the wintering grounds of the Whooping Cranes. On their migration route between Texas and Canada, such as in Oklahoma and Nebraska, the cranes face environmental destruction or changes of stopover habitat, such as water diversion into storage and use for irrigation, loss of river roosting habitat, and conversion of land to cropland.

While I was originally writing this section, the Rapid City newspaper reported on August 2, 2011, that the Whooping Cranes could "fall prey to wind projects" because the Fish and Wildlife Service was fast-tracking wind energy projects down the same corridor followed by Whooping Cranes and a number of other migratory birds. This overlap is yet another reason why it is a problem that almost all of the Whooping Cranes are in a single flock. Attempts to establish additional flocks in Wisconsin, Louisiana, and Florida aim to address this problem. These other flocks, however, are small and struggling so far.

Efforts to cross-foster Whooping Cranes with Sandhill Cranes have largely been unsuccessful even though the Whooping Cranes survived and migrated. Problems arose because the Whooping Cranes failed to mate with other Whooping Cranes, apparently having been imprinted on the Sandhill Cranes. A nonmigratory population of Whooping Cranes established in the mid-1990s near Kissimmee, Florida, has dwindled with high mortality and lack of reproduction.

The main Whooping Crane flock has had its own problems recently due to ongoing severe drought in Texas. Because little or no fresh water has been reaching the Whooping Cranes' wintering grounds due not only to the drought but also heavy water usage upriver, the coastal marshes where they winter have become increasingly salty. The high salinity levels kill many of the animals cranes eat, such as blue crabs, which resulted in crane deaths in 2010.

The Aransas Project (TAP, at http://thearansasproject.org/) is an alliance of citizens, businesses, organizations, and municipalities that want responsible water management of the Guadalupe River Basin in Texas, which will not only benefit persons along the waterways but also ensure that fresh water makes it to where the Whooping Cranes winter. TAP is also challenging new permits to withdraw water from the Guadalupe River, which feeds into the cranes' wintering grounds. In early 2014 in a lawsuit challenging the water management actions taken by the State of Texas, the initial decision in favor of TAP was appealed by the Texas Commission on Environmental Quality

(TCEQ). In December 2014 the Fifth Circuit Court of Appeals reversed the original decision that granted an injunction prohibiting the issuance of new water permits by TCEQ. At this writing, TAP is considering its remaining options in its pursuit of ensuring that water reaches the cranes on their winter range.

The Migratory Bird Treaty Act between the United States and Canada (1916) was instrumental in providing protection for Whooping Cranes and other migratory birds, as was the establishment of Wood Buffalo National Park (1922) and Aransas National Wildlife Refuge (1937).

The US Fish and Wildlife Service and the National Audubon Society established the Cooperative Whooping Crane Tracking Project in 1975 to obtain information on how best to manage the remaining crane populations and determine how to reverse the precipitous decline. Eggs obtained from wild nests allowed captive flocks to be established and birds to begin to be reintroduced into areas where they no longer were found.

Thanks to the Whooping Crane Eastern Partnership Reintroduction Project begun in 2001, Whooping Cranes have nested in central Wisconsin. The cranes were initially released into Wisconsin from captive flocks raised by handlers dressed to mask the human form to reduce imprinting on humans. After being transported to Wisconsin, the Whooping Cranes were conditioned to follow ultralight aircraft (http://www.operationmigration.org/), so they would migrate by following the planes to Florida wintering grounds. In addition, cranes have also been introduced into small groups with wild Whooping Cranes so the newcomers might learn the migration routes. The released birds are banded and outfitted with transmitters so that more can be learned about their migration routes, habitat, and other movements. As of May 2011, there were 105 surviving Whooping Cranes in this eastern population, and that number had decreased to 101 birds by the end of 2013 (http://www.bringbackthecranes.org/). An attempt is also being made to establish a second nonmigratory Whooping Crane population in the White Lake Wetlands Conservation Area in Louisiana.

You can learn more and become inspired about this magnificent bird by reading and marveling over the photographs in *Whooping Crane: Images from the Wild*, by Klaus Nigge.

A Personal Note: My first attempt to see a wild Whooping Crane on a trip to Texas was foiled by closure of a car rental company office there and the non-availability of any other rental cars. Although I was eventually able to see the

Florida Whooping Cranes that were being introduced there (not yet a self-sustaining population, so not "countable"), it wasn't until a couple of years later when we decided to move to Texas that I knew I'd finally have a chance to see countable Whooping Cranes.

For the eleven years that I lived in Texas, I saw Whooping Cranes at least once per year and sometimes much more often. While most of my sightings were through a telescope from the very high observation tower at Aransas National Wildlife Refuge, some years I was able to go out on one of the "whooping crane boats" that take visitors out to the refuge from Rockport-Fulton, Texas. There is very little that is quite as spectacular as seeing a grazing crane, or two or three of them, on the shoreline fifty feet from the bow of a boat, except perhaps to see them fly overhead as the boat travels down the canals near the refuge.

When I moved to South Dakota, I started making plans to look for a Whooping Crane as it rested on its migratory travels between Texas and Canada. In late March 2012, I had my South Dakota sighting of a Whooping Crane. Another person and I drove the roads in eastern South Dakota where a Whooping Crane had been reported the day before. The hunt was very reminiscent of the search for a needle in a haystack. Thousands upon thousands of Sandhill Cranes were spread out over miles and miles of rolling farmlands, primarily in fields of corn stubble. They spilled into ravines and onto hilltops, filling the air with their raucous calls and flying back and forth. Every now and then we would glimpse something white, but it was always Snow Geese. After more than eight hours of driving the country roads, we finally spotted a huge white bird flying low over some hills with about a hundred Sandhill Cranes. All of them landed out of sight in a valley, but we were able to drive to a spot where we could look into the valley and see the single Whooping Crane, a joy to behold.

SNOWY PLOVER AND PIPING PLOVER

Snowy Plovers and Piping Plovers are small shorebirds with pale brown backs and white undersides. They spend much of their time on sandy beaches and other barren or arid areas.

## *Snowy Plover (Charadrius nivosus)*

Global Population Estimate: from a worldwide total of about 370,000, the United States has about 21,000, a population that does not include the races found in Europe; the non-US races (Kentish Plover) have been recently separated from the Snowy Plover by the American Ornithologists' Union and are considered a separate species by this organization.

Snowy Plovers are small, pale shorebirds (just over six inches long) with a buff-gray back and white underparts. Like the similar Piping Plover, they have a dark patch over their white foreheads and a single breast band (the larger common Killdeer has two breast bands) that does not quite go all the way across the breast. Unlike the Piping Plover, the Snowy Plover has a thin, blackish beak, black legs, and a black cheek-patch. It nests on the ground in sandy coastal areas, as well as on some brackish inland lakeshores. It typically eats small invertebrates found in wet beach sand.

In the United States, the western race of Snowy Plover is found along the Pacific Ocean from southern Washington to the Gulf of California. There are also Snowy Plovers down the Pacific coast south of the United States to Chile. Inland birds are found in south-central Oregon down to the Salton Sea in California and Nevada, in Utah, Colorado, New Mexico, Oklahoma, Texas, and Mexico. The Gulf Coast race resides on undisturbed beaches around the Gulf of Mexico from Florida to Mexico. East of the US mainland, Snowy Plovers breed on Caribbean islands, the Bahamas, and islands off the north coast of Venezuela. Most birds of this species winter in the southern United States and into Mexico and Central America.

While detailed information about early Snowy Plover populations is sparse, with little data from the period before the 1970s, it is known that nesting Snowy Plovers were widely distributed throughout both the Pacific Coast and Gulf Coast regions before the twentieth century. It is estimated that since then there has been at least a 40 percent decline, with the number of nesting areas decreasing dramatically; however, new information in October 2014 indicated that there might be some recovery of Snowy Plover numbers, at

least in Washington State. The new world races of the Snowy Plover are variously classified as threatened or endangered by state (e.g., Oregon, California, Mississippi, and Kansas) and federal authorities.

As is the case with many beach-loving birds, Snowy Plovers have become trapped in discarded monofilament fishing line, which can cause the bird to lose its toes or feet and in some cases its life.

There is a recovery plan for the western Snowy Plover, and it includes increasing the numbers and productivity of Snowy Plovers and protecting the breeding and wintering plovers. As with many other shorebirds, Snowy Plover nesting areas on beaches, such as those in California and Florida, must be protected from disturbance and development by closing, fencing, or roping off beaches, banning pets, removing predators, and posting signs to educate the public. In some cases, periodically clearing invasive vegetation can yield conditions that approximate the natural state, as well as provide nesting habitat.

WesternSnowyPlover.org (http://www.westernsnowyplover.org/) is a coalition supported by public and private funding and devoted to the protection and recovery of the western Snowy Plover. This group provides information encouraging the public to learn about the plovers and how they can help save them and their habitat. The website provides an interactive map showing where volunteers can participate in public outreach by informing visitors about the threats to the Snowy Plover, helping to reduce disturbances by humans and pets, and helping to enhance and restore Snowy Plover habitat. A Personal Note: In addition to the treat of visiting the Texas Gulf coast to see Snowy Plovers each year I lived in that state, I was delighted to find Snowy Plovers during my exploration of Oklahoma in 2011. At the Great Salt Plains, which seems a very unlikely spot for a shorebird because it is an expanse of hot, windy, salty white sand that stretches as far as the eye can see and is set in the middle of a greater expanse of rolling prairie hills, I found them, spread out on the white sand, seeming to bake in the relentless sun. The birds apparently survive there by eating insects, which provide both their food and their moisture since the water on the plains is too salty. When the ground becomes too hot, they stand in the water.

### Piping Plover (Charadrius melodus)
Global Population Estimate: 6,410, all in the United States and Canada

The Piping Plover gets its name from its loud, distinctive whistle. Like the Snowy Plover, the Piping Plover is a small, sandy-colored shorebird that feeds and nests along the coast, where it can be seen running along, then suddenly stopping, and then starting again. The legs of the Piping Plover are much more colorful than those of the Snowy Plover, being bright orange, and the bill of the adult is orange with a black tip.

Unlike the Snowy Plover, which occurs both in North America and Europe, Piping Plovers are restricted to this continent, where they generally breed on sandy beaches along the Atlantic coast of North America, as well as along the shores of the Great Lakes and in the Midwest, on alkaline and freshwater lakes, reservoirs, rivers, industrial ponds, and gravel mines, foraging on the beaches and water edges for insects and crustaceans. The inland breeding range extends from alkaline wetlands in southeastern Alberta on the northern Great Plains to the north-central United States and southward along the main prairie rivers (Yellowstone, Missouri, Niobrara, Platte, and Loup) and watery areas in Colorado, Kansas, Montana, the Dakotas,

Nebraska, and Iowa. Along the Atlantic coast, breeding has been found from Canada (New Brunswick, Prince Edward Island, Nova Scotia, Quebec, and Newfoundland and Labrador) to Maine and south to North Carolina.

Piping Plovers winter on barrier island beaches, mudflats, and sandflats in the United States from North Carolina southward and on the entire Gulf coast, with most wintering in Texas. There are also important wintering areas in northern Mexico.

Piping Plovers were once found on much of the shoreline habitat east of the Rocky Mountains. Before the early 1900s, Piping Plovers were killed by hunters for feathers to decorate hats, which caused their first major population decrease. After that, coastal development and degradation, including efforts to stabilize beaches, have hastened the decrease in numbers.

The 1996 federal recovery plan for Piping Plovers established four recovery units (Atlantic Canada, New England, New York–New Jersey, and Southern) with specified criteria for increasing the number of breeding pairs in each unit so that the species could be removed from the endangered list. Similar goals were established for the Great Lakes region by a 2003 federal plan.

The US National Park Service is developing a management plan regarding the use of off-road vehicles on the Cape Hatteras National Seashore in

North Carolina, since interim protections have allowed the number of pairs of Piping Plovers to double in four years, with successful nesting occurring in places that long had been abandoned by the plovers.

There are also local conservation efforts aimed at preventing disturbance of Piping Plover nesting areas, such as closing portions of the beaches, putting barriers around nests to keep out predators, and regulating water levels so that nests are not flooded.

Piping Plover populations have increased substantially in some areas due to the conservation efforts that were made after the species was listed as endangered in 1986. In some areas, such as Cape Cod, Massachusetts, concerns have been expressed (by the FWS in December 2014) that the population rebound has been such that it is necessary to find ways that both people and the expanded Piping Plover population can coexist.

People can be careful not to walk and not to walk their dogs on coastal dunes, remembering to stay on boardwalks and marked trails. They can also participate in programs that survey, monitor, and protect beach-nesting birds and their habitats. Many coastal states have such programs for which people can volunteer, including Connecticut (http://www.ctaudubon.org/) and Delaware (http://www.dnrec.delaware.gov/fw/Volunteers/Pages/pipingplover.aspx).

A Personal Note: One of my favorite birding areas when I lived in North Carolina was the Outer Banks. This area is also loved by beachgoers, fishing enthusiasts, and those using off-road vehicles, and thus conflict was inevitable when beach access was cut off to protect nesting Piping Plovers. While these restrictions are seasonal and temporary, I have learned from my friends who still live in North Carolina that the intense difference of opinions on this issue has split the resident and vacationing community into opposing camps, sometimes escalating into violence. However, these measures in North Carolina have been very successful in enhancing nesting success for Piping Plovers, as well as other birds.

## *Mountain Plover (Charadrius montanus)*
Global Population Estimate: 8,500, all in North America

The Mountain Plover is of about the same size, shape, and tameness as the common Killdeer, but it is definitely not common. It also does not have the distinctive breast bands of the Killdeer.

In spite of its name, the Mountain Plover does not live in the mountains but rather is a bird of the Great Plains, being found on high, dry, level plains. Instead of running about on mowed cemetery lawns and gravel roads like the Killdeer, the Mountain Plover spends its breeding season on the remote short-grass prairies of the Great Plains, in Montana, Wyoming, and Colorado, and it winters on southern dirt and stubble or grassy agricultural fields from Texas to California, as well as in Mexico. Because Mountain Plovers do well on very short-grass prairies, heavily grazed lands and prairie-dog towns are good places to look for them.

The Mountain Plover, like many of its shorebird relatives, feeds by running across the flat land, stopping suddenly, looking for food, and repeating. It has been seen stamping a foot to cause prey to fly. It eats insects and other small arthropods, often accompanying cattle, as cowbirds do, to eat the insects that are disturbed by the cattle.

The Mountain Plover once was common on the Great Plains and had a huge range on the short-grass prairies in much of the West, from Montana to New Mexico and Texas. As settlers moved west and cleared the prairies to plant crops, the habitat required by Mountain Plovers decreased substantially. In addition, Mountain Plovers were hunted in the late 1800s. Because Mountain Plovers are difficult to find, both in their wintering grounds and on their breeding grounds due to the remoteness of their habitats and their cryptic coloration that blends with the surrounding grasses and dirt, accurate estimates of their populations, both now and in the past, are difficult.

Although breeding Mountain Plovers are found on plains and semidesert areas west of the short-grass prairie in Colorado, Montana, Wyoming, New Mexico, and the Oklahoma and Texas Panhandles, the Mountain Plover now primarily breeds in two counties in the entire country—Weld County in Colorado and Phillips County in Montana. The birds that breed outside this area are generally quite scattered and remote from each other.

Even dedicated bird surveys such as the annual federal Breeding Bird Surveys are unlikely to tell the whole story, but these surveys do show a decline in Mountain Plover numbers in the late twentieth century. Estimates are that the population has declined about 3 percent each year since the 1970s, so that the total is less than half of what it was then. Because Mountain Plovers have such a small population that continues to decline and because their range is so limited, there is concern about their future. As their habitat declines or becomes less suitable for Mountain Plovers, their population is expected to decrease further.

The Mountain Plover Festival is held annually in late April in Karval, Colorado, on the eastern plains of Colorado. This festival is aimed at bringing in birdwatchers, and it also allows visitors to learn how landowners and biologists can work together to preserve the nesting grounds of the Mountain Plover.

While there have been efforts to list the Mountain Plover as endangered, as recently as May 25, 2011, the US Fish and Wildlife Service announced that the Mountain Plover would not be listed under the Endangered Species Act.

Due to the difficulty of locating and then actually getting a good look at Mountain Plovers, particularly on their breeding grounds, overly enthusiastic searchers for Mountain Plovers might be tempted to get too near Mountain Plovers once they find an appropriate short-grass area. It is important that birdwatchers, listers, photographers, and others NOT enter the breeding areas of Mountain Plovers or closely approach them during breeding season. While the seemingly tame Mountain Plovers may not flush at the approach of the seeker, the close approach might cause them to abandon their nests.

A Personal Note: Almost every winter for the eleven years I lived in Texas, a birder's report on Texbirds (the listserv where people can post their Texas bird sightings) would announce that the Mountain Plovers had again arrived at their wintering grounds on the stubble and plowed fields northeast of Austin. Often the reports would give details as to the exact location on which country roads the plovers, as well as wintering longspurs, could be found. As soon as possible after the first birder's report, I would drive the two-plus hours to the designated Central Texas spots and begin an often fruitless countryside search, wandering up and down the gravel roads for hours and hours to try to find where the plovers had gone. Usually the wind across the flat fields was very strong and there were no birds to be seen anywhere, just vast fields of dirt and stubble and dry grass.

Some years, that was the extent of what I saw, but on the good trips I would see far out in a field a few tiny, whitish dots of equal size that didn't look as much like stones and pebbles as the other white dots in the field. Sometimes these dots could even be seen moving between and over the furrows of the plowed field. With the help of a telescope, I saw these white dots turn into Mountain Plovers. On rare occasions, with just the aid of binoculars, I could detect some shorebird shapes that would come quite close to the road, allowing the possibility of photographs. On the best trip of all during those years, a small flock of Mountain Plovers was sitting in a farmer's driveway not far from the road.

As I wandered around California in January 2008, it was good to find Mountain Plovers on a gravel road and on both sides of the road, seeming to pause in their foraging to pose for my pictures.

## *Wandering Tattler (Tringa incana)*

Global Population Estimate: 15,000; of this number, 10,000 are on the North American continent

Wandering Tattlers are shorebirds that are about an inch longer than an American Robin and slightly bigger than the similarly shaped Killdeer. Their breeding plumage is dark gray on the back and wings, with wavy dark gray bars on the birds' undersides and throats. Their bills are sharply pointed and about as long as their heads, and their legs are a dull yellow color. In winter the bars give way to a gray wash on their breasts and a white belly. As they walk along the shore or on rocks, Wandering Tattlers noticeably wag their tails and rear ends.

The breeding range of Wandering Tattlers is in Alaska and Yukon Territory, in Canada, and they winter on the coast of southern California and Mexico and in Hawaii. They are called "wandering" birds because they are so widespread on the Pacific shores of North America and Asia.

The status of Wandering Tattlers before the present time appears to be basically unknown.

Since 1918, Wandering Tattlers have been protected in the United States

and Canada under the Migratory Bird Treaty Act. It is difficult to determine, however, whether being protected has increased their numbers. The most likely threat to Wandering Tattlers is human activity on their wintering ranges because their breeding locations are generally quite remote. Information obtained from Christmas Bird Counts in California, where Wandering Tattlers winter, indicates that their numbers are decreasing, but Hawaiian numbers are remaining stable.

As with most other shorebirds, it is important to learn more about the population numbers and habitat requirements of Wandering Tattlers and to do something to identify and protect the areas that are important for the birds. Thus, monitoring of the tattler population and keeping their shoreline wintering habitats free from disturbance and predators are necessary.

Support for the National Audubon Society and its Important Bird Areas (IBAs) program (http://web4.audubon.org/bird/iba/), which is designed to identify and conserve important areas for birds, helps to protect areas that Wandering Tattlers need to survive. IBAs in Hawaii and Washington include areas where Wandering Tattlers winter.

A Personal Note: My only experience with Wandering Tattlers is in Alaska, which is the main place where they breed. I saw them in spring on Adak, an island in the Aleutian chain that extends toward Russia, as well as in Nome and in the Arctic National Wildlife Refuge on the North Slope. In each case, they were on the pebbly shores of a stream, somehow finding enough to eat in the icy water.

## Bristle-thighed Curlew *(Numenius tahitiensis)*

Global Population Estimate: 10,000, in Alaska and various Pacific islands

Bristle-thighed Curlews get their name from the stiff tiny feathers on their legs; however, these feathers are not noticeable enough to allow a birder to distinguish them from the other medium-large brown shorebirds that also have a long, downward-curved bill. Mostly you can identify Bristle-thighed Curlews by where you find them, since they are unlikely to be found anywhere but either the hilly inland tundra in western Alaska during breeding season, where they nest on the ground, or on some of the South Pacific atolls and islands, such as the Hawaiian Islands, in the winter. In Hawaii, the Bristle-thighed Curlew is known as the kioea. In Alaska they have an appearance similar to that of the Whimbrel, which may also be present and which

also has a dark brown eye-line and a brown cap on the head. The Bristle-thighed Curlew can be distinguished from the Whimbrel by the former's buff rump (as opposed to the Whimbrel's whitish rump) and a more buff-colored belly. In addition, the back of the Bristle-thighed Curlew is splotchier, with bigger buff spots on its back forming a coarser pattern than is seen on the Whimbrel.

The Bristle-thighed Curlew male defends a large territory and engages in an aerial display that can be seen if one takes a bird trip to the Nome, Alaska, area. After the breeding and nesting season is over, Bristle-thighed Curlews gather on meadows and tundra of the Yukon River delta on the west coast of Alaska and eat insects and berries to put on fat for their twenty-five-hundred-mile nonstop journey to the South Pacific. Once they arrive in the South Pacific they rest and eat and lose all their flight feathers when they molt (which other shorebirds do not do). The Bristle-thighed Curlew, which sometimes uses rocks to break open other birds' eggs to eat, also seems to be the only shorebird that uses tools.

Although it has been known to winter and was often hunted in the South Pacific from the days of James Cook's visit to Tahiti in the eighteenth century, the Bristle-thighed Curlew's nesting grounds were not known until the 1940s because of its remote nesting location. Collectors had found a few birds in Alaska in the late 1800s but presumed that they were vagrants that had wandered from their South Pacific home. Because its breeding grounds are so remote, the Bristle-thighed Curlew has fewer problems there than it does on its wintering grounds. On many South Pacific islands, introduced dogs and cats prey on the Bristle-thighed Curlews and can do much harm due to the birds' flightlessness for an extended period while they are on their wintering grounds.

Very little information exists on population trends of the Bristle-thighed Curlew. While a big part of the problem for Bristle-thighed Curlews is the fact that they become flightless on their wintering ground, nothing can be done about that. What needs to be done is to eliminate the threats to the places where these curlews winter, which include portions of the Hawaiian Islands National Wildlife Refuge. In their remote breeding grounds, the main problems are considered to be fires and damage to tundra caused by off-road vehicles such as those associated with mining.

It is difficult for most of us to provide any kind of hands-on help to such a remote species as the Bristle-thighed Curlew, but we can support funding for the national wildlife refuges, particularly in Hawaii, so that the bird's current habitat in Hawaii is protected and so that additional wetland habitat can be restored or reclaimed from abandoned agricultural and other lands. Additional research is also needed to better understand the habitat requirements and behavior of the Bristle-thighed Curlew.

A Personal Note: Nearly everyone who has seen a Bristle-thighed Curlew in continental North America has seen it in Alaska, on a particular hill on Kougarok Road outside of Nome, and this is also true for me. I have taken two birding trips to Nome, and in both cases we climbed this hill and eventually saw the curlew. The uneven terrain on the hill is composed of sturdy tufts of grass, each of which is surrounded by a lower area, making it very difficult to walk fast and coming with a high risk of turning an ankle unless you carefully walk in the lower areas around the tufts. This is difficult to do when you are in a hurry to keep up with the rest of the group so you can see the curlew if they flush it. Not to worry, however, because a Bristle-thighed Curlew is a big bird, and if it flushes, you are likely to see it. On my second trip there, not only did I see it fly up and display but with the other birders I watched it land again a bit farther ahead, where we could all see its long beak, head, neck, and shoulders above the grass and could compare it with the similar and slightly smaller Whimbrel that shared the hilltop with it.

## Long-billed Curlew (Numenius americanus)
Global Population Estimate: 20,000, all in North America

The Long-billed Curlew does not live in Alaska and therefore is not likely to ever be confused with the Bristle-thighed Curlew. Even if they did reside in the same place, there would be no confusion because Long-billed Curlews not only have a much longer decurved bill than the Bristle-thighed Curlew but also are larger, with longer legs, a longer neck, and cinnamon-colored underwings. The head of the Long-billed Curlew is also much plainer than that of the Bristle-thighed Curlew. The Long-billed Curlew uses flight displays during breeding season to attract a mate.

Long-billed Curlews use their long bills to capture beach creatures, such as shrimp and crabs, on the tidal flats where they winter and to excavate earthworms. They also eat a wide variety of insects.

The Long-billed Curlew breeds on grasslands from the north of Mexico all the way into the Canadian interior on open, flat, or rolling grasslands and prairies, with most of these birds wintering on mudflats and coastal areas in southern Texas and northeastern Mexico, as well as in California.

In the early period of settlement in the United States, Long-billed Curlews were heavily hunted as a food item. This is no longer legal. While the population of Long-billed Curlews is about twice that of the Bristle-thighed Curlew, it has suffered large population losses, especially in the Great Plains. It is

estimated that about a third of the historic breeding range of this species has been destroyed by agricultural expansion and urbanization.

Long-billed Curlews are variously designated as "highly imperiled" or "vulnerable" because of the small and decreasing size of the population and because there are habitat threats where they breed and where they winter.

There are local conservation groups concerned with protection of grasslands and with limiting public access to grasslands. To learn more so that you can assist or start such a group, the US Fish and Wildlife Service has put forth detailed management practices for the Long-billed Curlew (for example, see http://www.npwrc.usgs.gov/resource/literatr/grasbird/lbcu/lbcu.htm). Coastal and grassland area bird groups can be encouraged to support actions that will improve and protect habitat for Long-billed Curlews. Because Long-billed Curlews favor grasslands with relatively short grasses, light levels of grazing can improve their breeding success, as can periodic grassland burning. On their wintering grounds, such as southern rice fields, flooding of the fields attracts Long-billed Curlews.

## Red Knot *(Calidris canutus)*

Global Population Estimate: 1.1 million, of which about 400,000 are in North America

Red Knots are large shorebirds with short legs and a short bill. They are called "Red" Knots because of their very bright, rusty-red breeding plumage, which is unmistakable in breeding season. Red Knots include the most endangered population of Red Knots in North America, known as the "rufa" population, as well as populations that breed in northern Greenland and Russia and winter in Africa, Asia, and Australia. From about half to possibly more than 90 percent of the rufa population has a migration stop in the Delaware Bay area. The areas where they breed are typically hillsides with little vegetation. After breeding, the knots fly to coastal intertidal areas, where they feed on marine invertebrates.

The Red Knot, like the Buff-breasted Sandpiper, is a long-distance migrant, with the rufa population flying the more than ninety-three-hundred-mile trip from their breeding grounds in the middle and high Arctic regions to southern Chile and Argentina (Tierra del Fuego) and back each year. In the spring this migration is closely timed to coincide with the horseshoe crabs' egg-laying along the eastern coast of the United States, primarily along Delaware Bay and the Cape May peninsula. At these areas, Red Knots concentrate in huge numbers during migration and feed on the eggs of the

spawning horseshoe crabs, with possibly up to 90 percent of all rufa Red Knots being present at the same time on the Delaware Bay. The eggs are a rich source of protein fuel that supports the remainder of the Red Knots' flight to the Arctic.

Like Buff-breasted Sandpipers, Red Knots were hunted for market and for sport in the late 1800s and early 1900s. For example, many shipments of barrels, each barrel stuffed with something like sixty dozen knots, were sent from Cape Cod to Boston in the 1890s, with an estimated four thousand knots being killed in a single night by market hunters working by firelight. The species continued to decline, with the rufa population decreasing by more than 50 percent from the mid-1980s to 2003.

Because Red Knots found in the United States tend to concentrate together on migration and in the winter more than most other shorebirds, any disruption in or loss of habitat in their regular stopovers and wintering sites could be (and has been) more detrimental to their numbers than for other shorebirds. The rufa population of Red Knots that winters in Tierra del Fuego has declined dramatically, which is largely related to the increase in the 1990s in the taking of horseshoe crabs for bait in commercial fisheries, such as those that catch eels. The Red Knot was added to the list of federal candidate species in 2006. A final rule to list the rufa subspecies as threatened under the Endangered Species Act was published on December 11, 2014, with an effective date of January 12, 2015.

Since 1995, the population of the subspecies that breeds in the central Canadian Arctic region has gone from about 150,000 to the current level, estimated at about 18,000. Another subspecies that breeds in Alaska and is believed to winter on the Pacific coast of the United States and Mexico is not believed to have decreased so dramatically. In some of their wintering areas, Red Knot hunting for food or sport still occurs.

There is also concern that because Red Knots are so concentrated, the levels of genetic variation in the species are declining, which means that species resilience in the face of threats, such as to their habitat, could be sufficiently low so that a large proportion of the species could be in danger.

Identification of important Red Knot migration staging sites and protection of these sites from degradation and disturbance are important actions that need to be taken. A significant way to increase the rufa population of Red Knots would be to allow the Delaware Bay horseshoe crab population to increase so that their eggs would support a larger stopover population of Red Knots. Both Delaware and New Jersey have enacted regulations to limit harvesting of horseshoe crabs. Those who live in areas where horseshoe crabs and Red Knots are found should be vigilant so that such regulations are enforced and not relaxed. In addition, there are other stopover areas, such as on the Gulf coast of Florida, where habitat disturbance could be reduced. Disruption of migration habitats keeps the Red Knots from feeding, which limits fat accumulation and thus jeopardizes survival on the northern migration. Funding is needed to support research, surveys, and monitoring of the knots, as well as habitat conservation.

As mentioned herein with respect to Hudsonian Godwits, the Western Hemisphere Shorebird Reserve Network is working to identify key migration staging sites and wintering areas of Red Knots.

A Personal Note: Over the years, Red Knots have been a nemesis bird for me more than a few times. Knowing that they winter along the Texas Gulf coast and encouraged by reports that they had been seen recently by others, each year I would trudge down to some undeveloped or less-developed portion of the Texas coast and walk or drive endless miles along the sand looking for these large, squat, short-billed shorebirds. More often than not, I would return home to Fort Worth without having been able to see them. Generally, if I finally found them, there would be only a few, usually in their nondescript gray nonbreeding plumage. But every now and then, at the right time of the year, they would have some of their bright rusty ("red") breeding plumage. They usually would not fly as I approached, and I therefore was often able to photographically document the fact that I had finally found them. In early June 2008 I was able to see Red Knots in their full array of bright, rusty-red breeding plumage at Safety Lagoon outside of Nome, Alaska.

## *Buff-breasted Sandpiper (Calidris subruficollis)*
Global Population Estimate: 15,000, all in the Western Hemisphere

The Buff-breasted Sandpiper is a medium-small shorebird that is distinguishable from all the other buff or brown shorebirds by its unmarked buff underparts. All the other shorebirds have streaking, spots, or some other irregular marking. In addition, the rounded, dovelike head of the Buff-breasted Sandpiper is also mostly plain and buffy colored except for the striped crown and the big dark eye that stands out. The underwings of the Buff-breasted Sandpiper are

noticeably whiter than those of other shorebirds. The typical habitat of the Buff-breasted Sandpiper is also more likely to be short grass than the shore.

Buff-breasted Sandpipers breed in high Arctic areas, including the Arctic National Wildlife Refuge, tundra areas along barren ridges and creek banks, and other areas with little vegetation. During migration they are found in short-grass areas such as sod farms, golf courses, pastures, and airports. They winter in southern South America, migrating north and south along the center of the continents, crossing the Gulf of Mexico on the way to each destination. They feed primarily on insects.

Male Buff-breasted Sandpipers display solitarily before breeding but also gather and display in leks as discussed above with respect to prairie-chickens. As set forth in the account of this species in *Birds of North America Online* (*BNA Online*), this breeding behavior is unique among shorebirds of the Americas. Males defend these small territories and use them to display. After breeding, the females leave and make nests. As soon as the young hatch, they can find their own food and fend for themselves.

Buff-breasted Sandpipers were once numerous and were hunted for the market in a manner that seems analogous to that of the Passenger Pigeon. It is estimated that at the end of the 1800s there were possibly more than a million Buff-breasted Sandpipers. Uncontrolled hunting continued into the early twentieth century.

In addition to being substantially reduced by hunting, Buff-breasted Sandpipers also appear to be declining in numbers due to reductions in suitable habitat, although the actual cause is unclear. Other possible causes of the decline include airborne pollutants and pesticides that become concentrated in the eggs of the Buff-breasted Sandpipers.

Many South American countries, as well as the United States and Canada, list the Buff-breasted Sandpiper as a species of conservation concern.

Most of the possible actions to benefit the Buff-breasted Sandpiper relate to habitat improvement and preservation. Because this sandpiper traverses so many countries it is important that there be coordinated international efforts. Thus, the Convention on the Conservation of Migratory Species of Wild Animals (known as the Bonn Convention), of which 120 countries are members, should be supported by all these countries and should be joined by the United States and Canada (which had not done so as of May 1, 2014).

Elected officials should be encouraged to support not only the Bonn Convention but also the preservation of the Arctic National Wildlife Refuge, since it is the breeding site of ALL of the Buff-breasted Sandpipers in the United States. The organizations working toward this end include Defenders of Wildlife (http://www.defenders.org/) and the Natural Resources Defense Council (http://www.nrdc.org/).

A Personal Note: Although "cute" is probably an overused word with respect to birds, Buff-breasted Sandpipers in my opinion are the cutest of the sandpipers, with their round tan face, bright beady eyes, and overall round appearance.

When I lived in North Carolina, one of the nearby places reliably featuring Buff-breasted Sandpipers on migration was a sod-farm area in South Carolina. Each spring, reports would come that shorebirds had arrived at the sod farms, and birders would troop down there to drive up and down the roads between the lawnlike fields, scanning the wet, bright green grass for shorebirds. Finding a few Buff-breasted Sandpipers was usually the high point of these trips.

After I moved to Texas, on a memorable Fort Worth Audubon field trip to a lake north of the city, as we were walking across an open area of wet sand and widely dispersed grass and other plants, we were delighted to suddenly be in the midst of a small flock of scurrying Buff-breasted Sandpipers. For some reason, perhaps due to their being tired after a stretch of migrating, they seemed reluctant to fly and allowed all of us to get close views.

---

KITTLITZ'S, XANTUS'S (NOW SPLIT INTO SCRIPPS'S MURRELET AND GUADALUPE MURRELET), AND CRAVERI'S MURRELETS

Kittlitz's Murrelet, Xantus's Murrelet (since August 2012, Scripps's and Guadalupe Murrelets), and Craveri's Murrelet are all robin-sized, football-shaped, generally black-and-white oceanic species that only come to land to nest.

## *Kittlitz's Murrelet (Brachyramphus brevirostris)*

Global Population Estimate: 24,000, of which as many as 17,000 are on the North American continent

Kittlitz's Murrelets are small diving birds that occur in near-shore and ocean waters off Alaska and eastern Siberia, with the population centered on the Bering Sea. They are one of the rarest seabirds in North America. Instead of

colonially nesting on islands, as do most seabirds, Kittlitz's Murrelets nest above the timberline on remote inland mountaintops, such as those along the coasts of Alaska and Siberia, with the nests on the bare ground, often near snowfields. The color of Kittlitz's Murrelets during breeding season is gray-ish brown (the color of dirt), but in winter they become black and white, like many other seabirds, including the other murrelets discussed in this book.

Kittlitz's Murrelets mainly feed close to shore, often near glaciers that reach the ocean waters. They eat small fish, krill, and other zooplankton. The young murrelets are fed fish.

Due to the remoteness of the Kittlitz's Murrelet habitat, little is known of the size of their earlier populations, and in fact, amazingly little is known of current populations. It is known, however, that Kittlitz's Murrelet populations are decreasing precipitously. Estimates suggest an 18 percent annual decline since 1989 in some areas and an 80 percent decline in others, with an estimated drop to 1 percent of its 2000 level by 2026. It is estimated that the Alaskan population of Kittlitz's Murrelets has decreased by nearly 90 percent in the past fifteen years due to glacial water habitat loss. At the current rate of decline of some populations of Kittlitz's Murrelets, they will disappear in thirty to forty years.

Various conservation measures, such as trying to list the Kittlitz's Murrelet under the Endangered Species Act and drawing up guidelines to avoid disturbing them, are important. There is also concern about gill netting, which needs to be reduced. Since so little is known about Kittlitz's Murrelets, it is also important to identify and protect both breeding and nonbreeding areas. Also, it is critical that research be done to determine the actual population trends and the factors affecting their numbers.

The future of the Kittlitz's Murrelet seems directly related to global warming, so unless the trend is reversed there may be little that can be done for the Kittlitz's Murrelet.

A Personal Note: I have very little experience with any of the three tiny murrelets covered in this book, since their ocean habitat is so remote from where I normally go birding. During 2008, however, when I decided to visit Adak on the Alaskan Aleutian chain, I did have the opportunity to see Kittlitz's Murrelets. On May 11, to my great delight, in the very bird-rich Clam Lagoon there were six Kittlitz's Murrelets in winter plumage.

## Scripps's Murrelet (Synthliboramphus scrippsi) and Guadalupe Murrelet (Synthliboramphus hypoleucus), previously one species, Xantus's Murrelet

Global Population Estimate: approximately 5,600 Xantus's Murrelets (species was split in 2012); some sources estimate substantially more birds, all of which occur off the west coasts of California and Mexico

The very similar Scripps's Murrelet and Guadalupe Murrelet species together are also one of the world's rarest seabirds. Like many tiny rounded seabirds (nine and three-quarters inches long, or smaller than a robin), Scripps's and

Guadalupe Murrelets are black on top and white on the chin, throat, and belly. Each of these species has a white eye-ring, white underwings (distinguishing them from Craveri's Murrelet, below), and a short, stout bill. They are rarely seen from the coast as they prefer the deeper, warm offshore Pacific waters beyond the continental shelf. They feed, usually in pairs, by diving and swimming underwater after fish and other small sea creatures.

Scripps's Murrelets breed on islands off the coast of California (Channel Islands) and Baja California, Mexico. Guadalupe Murrelets breed from Guadalupe Island south to the San Benito Islands of Mexico. Their nests are on cliffs, canyons, and steep areas on the rocky islands, usually concealed by vegetation. Once the young hatch they are quickly on their own, with the parents flying off to sea after a few days and the chicks making their own way to the sea, often over rocks and down steep areas, sometimes leaping off cliffs two hundred feet above the water below. Amazingly, once the chicks are down in the water, they get back together with their parents, who care for them there for a few months. After breeding season, they move farther out to sea, usually staying in the south but sometimes going as far north as British Columbia.

Because all US breeding colonies of these two species in California are in national parks, private protected areas, or military bases, they are legally protected, so it should be possible to reduce or eliminate predator problems. However, many Mexican breeding colonies are not protected. Even areas that are protected need to be monitored to be sure that the protection continues and is effective.

A Personal Note: Because one must go far out in the Pacific Ocean to see Scripps's Murrelets and Guadalupe Murrelets, and because they are so rare, I can remember seeing only one of them, ever. That sighting was in November 2008 on a pelagic trip out of San Diego and was a delight to even the experienced oceangoers on board, as it was one of the rarer Guadalupe Murrelets (then called the "hypoleucus" race), which has a bit more white on its cheek than the more usual Scripps's Murrelets. A very pleasing attribute as far as I was concerned was that the Guadalupe Murrelet did not bolt away across the ocean surface right away as many seabirds do but actually allowed us to see it and to photograph it.

## Craveri's Murrelet *(Synthliboramphus craveri)*

Global Population Estimate: 15,000 to 20,000, all of which occur off the west coasts of California and Mexico

Craveri's Murrelets are even smaller (about eight and a half inches long) than Scripps's and Guadalupe Murrelets and typically are found farther south than those species. While superficially very similar in their black-and-white appearance, a closer inspection reveals that the Craveri's Murrelets have more black under their eyes and chin and also have a black partial collar that the newly specified murrelets do not have. When they fly, instead of having white underwings, Craveri's Murrelets have mottled gray underwings. Their bill is somewhat longer and thinner than that of Scripps's and Guadalupe Murrelets, but when out on the ocean the bird must be very close for that distinction to be made. They also dive down into the water to catch prey, swimming along beneath the surface after fish.

Craveri's Murrelets breed on arid Gulf of California islands in the coastal waters of Mexico and can be found either on these islands or in the surrounding waters. Their eggs are usually laid directly on bare rock or in burrows or rock cavities. They feed on larval fish, such as rockfish and herring. After they are through breeding, they wander farther out to sea both to the north and south of their breeding islands.

I am unaware of any information that is available to help us understand how the earlier population differed from the current population of Craveri's Murrelets.

As with other island-nesting birds, introduced predators such as cats and rats are the main problem for Craveri's Murrelets. In addition, oil spills from tanker ships are a major threat, as are drift gill nets (which may drown the birds) and disturbance of their nesting areas.

A biosphere reserve has been designated under the UNESCO Man and the Biosphere Program. Called Islas del Golfo de California, it covers 124 islands in the Gulf of California that are very important for Craveri's Murrelets. Within this reserve, introduced mammals have been or are being eradicated so that the Craveri's Murrelets can nest. In addition, management plans are being developed for all the islands where Craveri's Murrelets are known to breed, and the nesting colonies are being monitored and a determination made as to what other measures will be of benefit to the birds. Eradication of non-native mammals from breeding islands is most critical, as is control of human disturbance. In many cases, there are no regulations to keep humans, such as tourists, from engaging in behaviors that are detrimental to Craveri's Murrelets on the nesting islands. Increased regulation, environmental education, and signs identifying fragile areas, as well as enforcement of current regulations, are all necessary.

One of the best ways to help Craveri's Murrelets, as well as other seabirds, is to become better informed of what is being done for them and to support organizations that are doing something to help. One such organization is Island Conservation (http://www.islandconservation.org), which is a non-profit organization with a focus on restoring the native seabirds and island ecosystems of Baja California and southern California. Gifts to Island Conservation go directly toward work on the islands to remove predators and restore the habitat needed by Craveri's Murrelets.

A Personal Note: Due to the combination of the rarity of Craveri's Murrelets, their range being in Mexico, and my lack of having taken very many southern California pelagic trips, I have not seen a Craveri's Murrelet and therefore have no personal note to share. I do hope this situation can be remedied someday.

### *Ivory Gull (Pagophila eburnea)*
Global Population Estimate: 15,550 to 23,900, of which about 600 breed on the North American continent

Ivory Gulls are small circumpolar Arctic gulls, being found in Canada, the United States, Greenland, and Norway. While most other species of gulls are basically white when viewed from a great distance, if you see them at closer viewing distances almost all of them have gray backs with darker or black wing-tips. A notable exception to typical gull coloration is the adult Ivory Gull, which is pure white ("ivory") except for its grayish bill, which has a yellowish orange tip, and its black eyes and feet. During their first winter, Ivory Gulls have blackish spots on the tips of their wing feathers and a black tip on their tails. While Ivory Gulls are birds of the far north, periodically there is great excitement when one wanders south of Canada and is found by a birder (e.g., in Quincy, Illinois, in January 2015).

Ivory Gulls breed near pack ice, placing large nests on rocky or cliff areas, with the only breeding area in Canada being in Nunavut Territory. They winter on ice out in the northwestern Atlantic Ocean, such as the waters north of Newfoundland. They eat fish and other small ocean animals, as well as carrion and seal placentas.

It is known that the number of Ivory Gulls has declined since the founding of the United States, even though the species was seen and recognized as unique in the early 1600s. There are indications that this decline is accelerating in some areas; however, because of the remoteness of where they breed and winter, information is quite incomplete.

While the proportion of the global population of Ivory Gulls that breeds in the United States is small, it is still important that in the Canadian Arctic the number of Ivory Gulls has decreased by 80 to 85 percent since the early 1980s, and many breeding colonies have disappeared. It is estimated that at this rate there will only be 190 breeding Ivory Gulls in Canada by 2016. Although it is believed that much of this decrease is due to climate change, there is some shooting of Ivory Gulls for food in northwestern Alaska. Most of the Ivory Gull breeding colonies are not near human communities, so most are safe from hunting. In some areas, the eggs of Ivory Gulls have been found to be contaminated by high levels of mercury and certain organochlorines (as is the case with some other seabirds).

As with many birds that live on or near the ocean, Ivory Gulls are affected by ongoing water level increases and changes in ice amounts due to global warming. Therefore, efforts to limit greenhouse gas emissions, though seemingly abstract, are concrete efforts that can benefit Ivory Gulls. In addition, the threat of pollutants, such as mercury, in eastern Canada needs to be addressed.

The Ivory Gull's international status means that it is important that there be treaties recognizing the in-trouble status of Ivory Gulls and requiring northern countries to preserve and protect the Ivory Gull habitat. Conservation of Arctic Flora and Fauna (CAFF, http://www.caff.is/), which is a working group of the Arctic Council, has produced the International Ivory Gull Conservation Strategy and Action Plan to facilitate circumpolar implementation of initiatives designed to conserve and protect Ivory Gulls. It includes twenty specific action items addressing the goals of ensuring that nonconsumptive uses of Ivory Gulls do not threaten their populations, minimizing the adverse effects of commercial activities, protecting key habitat, ensuring proper coordination with existing and planned conservation programs, encouraging awareness of the CAFF strategy, and providing reliable information about Ivory Gulls to implement the strategy and conserve Ivory Gulls globally.

A Personal Note: My only sightings of Ivory Gulls were in 2008 in Alaska. On June 4, when I was birding in Gambell on St. Lawrence Island off the west coast of Alaska, an Ivory Gull (or more than one) coasted by multiple times when we were staring out to sea looking for seabirds. Once one of them actually landed on the water and pecked at it briefly, a couple of hundred feet or less off the beach where we sat, freezing. On October 11 and 12, sometime in the middle of the day, two adult Ivory Gulls came by the north coast of Alaska when I was birding in Barrow. We also saw a first winter Ivory Gull there with its spotted back and wing-tips.

## *Flammulated Owl (Psiloscops flammeolus)*
Global Population Estimate: 37,000, all in North America

Flammulated Owls breed in the western mountains (Sierra Nevada, Cascade, and Rocky Mountain ranges) from southern British Columbia to northern Mexico and winter from Mexico to southern Guatemala. They are small owls, about six and three-quarters inches long, and weigh only about two ounces. The word *flammulated* in their name refers to the reddish-brown, flamelike coloration on their shoulder feathers. Otherwise their plumage is mainly either an overall gray or reddish-gray coloration. In either case, the mottled patterning helps them to blend in with the bark of the trees on which they roost. They have very short ear-tufts that are not always easily visible.

Flammulated Owls make a very low, soft, hooting vocalization that sounds like it is made by a much larger owl. They call only at night and also only hunt at night for moths, crickets, grasshoppers, beetles, spiders, and centipedes.

The strictly nocturnal and secretive habits of Flammulated Owls, as well as their scattered mountain habitat, mean that little is known of their population numbers or distribution, their ecology, and their overall status. The mountain forests most frequented by Flammulated Owls are composed of ponderosa pines mixed with oaks or aspens, with shrubby undergrowth. Their breeding areas typically are open coniferous forests with some old trees, clearings, and thickets. They nest in holes in trees originally excavated by woodpeckers such as Northern Flickers or Pileated Woodpeckers and sometimes in nest boxes.

Due to their nocturnal nature and ability to hide, it is very difficult to know the current status of Flammulated Owls and even more difficult to know whether their populations have decreased as human populations have expanded into the owls' territories.

Because Flammulated Owls wander extensively, conclusions that they are quite common are likely inaccurate. Because of their very low reproductive rate, there is concern that the population is particularly vulnerable and would have difficulty rebounding if the population declines, for example, due to environmental degradation or pollution.

There may be little that most of us can do directly to help the Flammulated Owl unless we own property where they might be found. Protection of habitat suitable for Flammulated Owls, particularly minimizing removal of dead trees or branches where the owls might nest, is important. It may be possible to reduce the effects of logging by providing artificial nest boxes. The Roadless Area Conservation Rule in US Forest Service territories was designed to protect the habitat of birds such as the Flammulated Owl. After public comments on this rule were submitted, the rule was challenged in court. After the roadless rule survived its final legal challenge by the state of Alaska, it became law on March 25, 2013, with the district court holding that no further challenges were allowed nationwide because of the statute of limitations.

Use of insecticides that kill moths, which are important prey of Flammulated Owls, especially in breeding season, can affect reproductive success of the owls and should be reduced or eliminated in areas where the owls are found.

As with other hard-to-find species that are actually or potentially in danger, it is very important that sightings of Flammulated Owls be reported, for example, on eBird, and that additional studies be done so that the number of Flammulated Owls can be more reliably estimated. Hawk Watch International has sponsored a banding study of Flammulated Owls in the Manzano Mountains of New Mexico to increase the knowledge of this species, and it does important work in monitoring a number of raptors (http://www.hawk-watch.org/support).

A Personal Note: I have seen only one Flammulated Owl, although I have heard them a couple of times. All of these Flammulated Owls were in the Chisos Mountains of Big Bend National Park in southwestern Texas in the summertime. To get to this area where such an observation is even possible, one must hike about four and a half miles uphill on a stair-step climb to the Boot Springs area. Because Flammulated Owls call only in the dark, one must be high in the mountains in the middle of the night, ideally having brought your tent up the mountain so you have someplace close to sleep after you've seen (or heard) the owls. All of that I've done, and sometimes I have been rewarded by hearing the owl. The one time that I saw one was by flashlight on an overhead branch, and it was a very welcome sighting.

## Spotted Owl (Strix occidentalis)
Global Population Estimate: 15,000 (estimated in 2000), all in North America

Spotted Owls are large (sixteen to nineteen inches long, with a three-and-a-half-foot wingspan) brown-and-white owls of dense wooded areas, including dense forests and wooded canyons. They nest and roost in old-growth forests and eat prey that also lives in these forests. Unlike most other owls, Spotted Owls (and the related Barred Owls) have dark, not yellow, eyes. One major difference between the rare Spotted Owls and the much more common Barred Owls can be learned from their names—Spotted Owls have spots over most of their bodies, while Barred Owls are barred underneath with vertically streaked breast feathers.

Spotted Owls eat small mammals such as flying squirrels, wood rats, mice, bats, and voles, as well as small birds, reptiles, amphibians, and insects. They nest in holes in trees, on broken-off trees, on platforms or old hawk nests, or on cliffs. Before getting their feathers the young are large, white, and fuzzy-downy, with black eyes.

The Spotted Owl species is made up of three subspecies—the Northern

Spotted Owl (southern British Columbia to northern California), the California Spotted Owl (California), and the Mexican Spotted Owl (Utah and Colorado to Texas and Mexico). All three subspecies are considered to be in trouble under one official categorization or another. Because of the fact that Spotted Owls, particularly Northern Spotted Owls, require old-growth mature forest, usually coniferous forest, Spotted Owls have been in the center of controversy over clear-cut logging of their habitat for many years. Areas such as national forests that generally have not been as extensively logged have most of the remaining concentrations of Spotted Owls.

The Spotted Owl population in the relatively undisturbed old-growth forests of the early days after European settlement was undoubtedly substantially greater than it is today; however, I have been unable to determine the extent of decline. It is likely that very little information on the population size was known in the early days of our country.

The current Spotted Owl populations of the three subspecies were minimally estimated in Birds of North America Online as of 1993 as follows: Northern Spotted Owl, 3,778 pairs and 1,001 single territorial owls; California Spotted Owl, 3,050 individuals; and Mexican Spotted Owl, 777 to 1,554 owls in the United States and 38 in Mexico. In areas where the owls reside and there are serious disturbances, such as logging of their nest trees, there is strong evidence to indicate that their numbers are rapidly declining.

The Spotted Owl debate has been very impassioned and public, and there have been many milestones and decisions. In 1936, eighteen years after enactment of the Migratory Bird Treaty Act, the Spotted Owl was designated a protected migratory bird. It wasn't until the 1970s, however, that guidelines were developed for Spotted Owl protection. In 1973 the Endangered Species Act was passed, but it was not until 1989 that the US Fish and Wildlife Service proposed that the Spotted Owl be listed as threatened. After much publicity and furor, the Spotted Owl was officially listed as threatened in 1990; however, the California Spotted Owl has not been officially listed as threatened. The case continued, however, with litigation and executive pronouncements of various kinds. While logging has been reduced in some areas and certain areas have been declared critical habitat for Spotted Owls, economic hardship resulted where logging supported communities, which has led to changes and counterchanges in plans for Spotted Owl recovery under different administrations.

One of the best things for citizens to do is to keep informed so that changes are not made in the laws that make the situation even worse for Spotted Owls.

The Northwest Forest Plan (1994), a critical Northern Spotted Owl conservation plan covering 24.5 million acres over three states, coordinates forest management across nineteen national forests, seven Bureau of Land Management districts, six national parks, and eight federal agencies.

The Endangered Species Act encourages individual conservation plans by land managers. A Habitat Conservation Plan (HCP) is a long-term plan (fifty to one hundred years) to protect threatened and endangered species and their habitats. HCPs have been developed, for example, on nearly two million acres in Washington State on state and private lands to protect the Northern Spotted Owl.

In June 2011, the US Fish and Wildlife Service released the Revised Recovery Plan for the Northern Spotted Owl, which includes thirty-three recommended actions to achieve recovery of the owls (www.fws.gov/oregonfwo). The plan was revised in 2012 to take into account the growing problems with Barred Owls and to establish a plan for determining critical habitat for the Northern Spotted Owl. This plan provides guidance, however, and is not regulatory. It establishes an executive group of federal, state, and tribal executives and a team to oversee implementation of the plan. Because of increasing evidence that competition from Barred Owls is a major factor in the decline of Spotted Owls, federal plans were announced in February 2012 for a four-year experiment to shoot Barred Owls in habitat considered critical for the survival of Spotted Owls in Washington, Oregon, and northern California. The plan also provides for nonlethal removal of Barred Owls by capturing and relocating them or placing them in permanent captivity. A final revised critical habitat rule for Spotted Owls was issued by the FWS in late 2012, which was followed by a number of lawsuits challenging critical Spotted Owl habitat designated under this rule.

Leading organizations involved in protecting Spotted Owls and their habitat include the Seattle Audubon Society (www.seattleaudubon.org) and the Washington Forest Law Center (www.wflc.org).

A Personal Note: In 2003 I learned that Spotted Owls could be found in the Guadalupe Mountains of Texas, much closer to my home than the traditional spots in Arizona where most birders see their first Spotted Owl. When I followed a friend's instructions for getting to one of the best places to find Spotted Owls in Texas, I also learned that a big Spotted Owl is amazingly difficult to see even when it is sitting out in the open on a bare tree branch. Its color and its spots fade into the background branches as if the owl was not even there. Therefore, unless one is really lucky, it requires scanning every branch, over and over, to see a Spotted Owl. Also, Spotted Owls rarely fly during the daytime but sit unmoving when people walk around or under them.

My most memorable Spotted Owl experiences were in 2006 and 2008. In 2006 I hiked up into the Guadalupe Mountains to get what I was hoping would be my annual view of a Spotted Owl. As is usual in the Texas mountains during summer, the air and I became hotter and hotter as I clambered over and around rocks on the sporadically marked trail. When I reached the canyon area where rock walls towered high above the gently sloping terrain and clusters of relatively large trees huddled along the trail mostly in the shade of the canyon walls, I began to check the tree branches for owls. I saw no owls there (I had never seen owls at that part of the trail), and I went farther up the trail to the more likely spots for the owls. When I reached a "T," I investigated all the branches of trees in both the right-hand canyon (where I'd usually found an owl in previous years) and the left-hand canyon. I could not find any owls. Few birds of any kind were evident, except I could hear the sounds of Canyon Wrens and Cordilleran Flycatchers.

Sad and disappointed, I retraced my steps down the main canyon, re-scanning every branch just in case. When I reached a particularly broad area between the canyon walls, where there were many trees on my left and it was very quiet in the shelter of the canyon, I slowed down and looked around again. Suddenly, I heard grass rustling in the brush under the trees, and my heart stopped. I had been told that in the Texas mountains one was generally always in the territory of a mountain lion. A mountain lion—in a narrow canyon—with me? I slowly inched away from the trees, looking everywhere, trying to figure out what I had heard, and frantically trying to decide what I would do if a lion materialized. Then I saw an owl, a large adult Spotted Owl, dozing on a branch of one of the trees. Clearly, it was not what had not made the rustling sound, but I forgot about the sound in the delight of seeing and photographing the owl. Remembering the sound, though, I again began backing out and then saw a second Spotted Owl, a couple of trees away from the first one. I had never seen two of them in the same year, much less the same day and place. And then I saw a large, whitish, fuzzy blob in a third tree, almost as big as the feathered adults, its black eyes staring down at me—a young Spotted Owl. More pictures of course. And then I heard the sound again—and saw a second fuzzy baby owl, on the ground, noisily bull-dozing its way through the grass under the trees. Mystery solved, and a spectacular total of four Spotted Owls seen and photographed and forever in my memory. It was one of my best days of birding ever.

Although I again saw Spotted Owls in 2008 in Arizona, it took many trips to find them, finally, in Miller Canyon. Seeing them requires substantial effort, going to one of the few places they are, and lots of luck!

## *Red-cockaded Woodpecker (Picoides borealis)*
Global Population Estimate: 20,000, all in North America

The Red-cockaded Woodpecker gets its name because of tiny, nearly invisible red tufts on the sides of the male's head that are usually hard to see because they are covered by other feathers. Rather than look for the red "cockade," birders would do better identifying field marks such as the big white cheek-patch on this small black-and-white woodpecker (eight and a half inches long). The black and white areas are in the form of horizontal stripes on its back, a black cap, and black surrounding the white cheek-patch.

Red-cockaded Woodpeckers, like most other woodpeckers, eat insects such as beetles, cockroaches, caterpillars, and wood-boring insects, as well as spiders and sometimes plant fruit and berries. They live in mature pine forests, preferably longleaf pines or other southern pines. Unlike other wood-peckers, Red-cockaded Woodpeckers do not hollow out dead trees or nest in holes made by other animals or birds but undertake the difficult task of digging a hole in a living pine tree. The preferred pine trees for this operation are those with a fungal condition called "red heart rot" that causes the center of the tree to become soft. Even then, it takes one to three years for Red-cockaded Woodpeckers to excavate a cavity. It is easy to find a tree where a

Red-cockaded Woodpecker cavity is being hollowed out or has been formed because the living tree exudes sap around the excavation, forming waxy "candle trees." It is believed that the waxy layer below the hole in the pine tree deters, and possibly often prevents, large snakes from entering the hole.

Unlike many other woodpeckers, Red-cockaded Woodpeckers live in groups or colonies (the colonies or the nesting trees of the colonies are called "clusters") in a defined territory, with one to twenty cavity trees in the colony's territory of three to sixty acres. The colonies contain usually four to nine or more members, with the members being one breeding pair and "helpers," which are usually males raised during the previous breeding season who help incubate the eggs of the breeding pair and feed the young. Red-cockaded Woodpeckers often have the same mate year after year, and they nest in the roost cavity of the male. The eggs are incubated by members of the colony. When the young leave the nest, they join their parents and the others in the colony, but when grown, the young females usually leave the colony to find a mate. In the winter, multiple Red-cockaded Woodpeckers apparently sometimes roost in the same hole.

The holes made by Red-cockaded Woodpeckers are often later used by other woodpeckers and other hole-nesting birds such as Carolina Chickadees, Eastern Bluebirds, and Tufted Titmice, as well as by flying squirrels and other animals.

The range of the Red-cockaded Woodpecker once extended from Florida north to New Jersey, west to eastern Texas and Oklahoma, and inland to Kentucky and Missouri. It is believed that there were once at least one hundred times as many Red-cockaded Woodpeckers as there are now, an estimated 1.0 million to 1.6 million groups, a "group" being the family unit of Red-cockaded Woodpeckers.

One estimate of the current number of Red-cockaded Woodpeckers is about 5,000 colonies containing about 12,500 birds (or possibly a few thousand more) in some of the remaining pine woods habitats from Florida to Virginia and west to Oklahoma and eastern Texas. Red-cockaded Woodpeckers are currently found in eleven states, on federal, state, and private lands. They are no longer found in New Jersey, Maryland, or Missouri. Even before the Endangered Species Act went into effect, efforts were being made to manage and preserve Red-cockaded Woodpecker habitat.

Because many otherwise possibly suitable habitats do not contain large enough trees for the Red-cockaded Woodpecker nests, artificial nesting cavities have been created in smaller longleaf pine trees. Either a cavity is drilled or, preferably, a nesting cavity is carved out and a manufactured nest is inserted in the cavity. These constructed cavities actually are used by Red-cockaded Woodpeckers when there are insufficient cavities in a colony's area, rather than the colony moving away to find a better area. The constructed cavities also enable the establishment of new groups of Red-cockaded Woodpeckers. In addition, restrictor plates have been used around the holes to keep other animals from enlarging and using active Red-cockaded Woodpecker nest holes.

An additional program of moving young female Red-cockaded Woodpeckers to other isolated Red-cockaded Woodpecker groups that are lacking females both reduces extinction of the isolated groups and allows maintenance of genetic diversity in the isolated groups.

The US Fish and Wildlife Service Red-cockaded Woodpecker Recovery Program is aimed at conserving these woodpeckers and their habitats. There are management plans for federal and state agencies on lands where there are Red-cockaded Woodpeckers. Even on private lands, nearly half of the Red-cockaded Woodpeckers benefit from various memorandums of agreement and other covenants, as well as from Habitat Conservation Plans.

Special management guidelines for Red-cockaded Woodpeckers on national forests and military installations have been developed by the US Forest Service and the Department of Defense, resulting in population increases by as much as 50 percent at military installations in Florida (Eglin Air Force Base), Georgia (Fort Benning and Fort Stewart), North Carolina (Fort Bragg and Camp Lejeune), and Louisiana (Fort Polk).

These intensive efforts have enabled a small comeback in Red-cockaded Woodpecker populations, with one source indicating that the number of active clusters increased from about 4,700 in 1993 to about 6,105 in 2006. Local efforts, such as one undertaken by the Nature Conservancy at Piney Grove Preserve in Virginia, have also resulted in increases, with a report in January 2015 of a tripling of the number of birds from 2001 to 2014.

The North Carolina Sandhills Conservation Partnership (NCSCP, at www.ncscp.org) was formed in 2000 and has as its mission the coordination of development and implementation of conservation strategies for the Red-cockaded Woodpecker, its habitat, and organisms that share its habitat. The steering committee members of the NCSCP represent various federal, state, and nonprofit conservation groups. The partnership specifically seeks input

from its stakeholder groups, which include The Nature Conservancy (www
.nature.org), the Sandhills Area Land Trust (www.sandhillslandtrust.org),
and the Sandhills Ecological Institute (www.sandhillsecological.org), as well
as federal and state entities.

A Personal Note: A nice thing about Red-cockaded Woodpeckers is that they
do not migrate and can be found all year in the same piney woods, unless
something has happened to their colony. A bad thing, however, is that they
rise before dawn and leave their nesting/roosting holes, rarely to be seen
again until just before dusk, when they return from their day of feeding.
Unless you know which holes they went to roost in the night before, you
might spend dawn after dawn staring at abandoned holes in pine trees from
which no birds come forth. During the day, when the foraging flock can
wander far from the tree holes, it is often difficult to find any of them, even in
areas where they nest and roost.

One of the highlights of birding when I lived in North Carolina was going
toward the coast to the piney woods habitats where the Red-cockaded Woodpeckers resided. One of my favorite places to look for Red-cockaded Woodpeckers was the Millis Road Savanna in the Croatan National Forest, where
other woodpeckers of the Southeast (Red-headed, Downy, and Hairy Woodpeckers and Northern Flickers) could also often be found, as could Bachman's
Sparrows and Brown-headed Nuthatches. The savanna is burned often, which
keeps the pine and grassy habitat in the prime, open condition preferred by
Red-cockaded Woodpeckers. Sometimes when I visited, I was lucky enough
to be in the center of a foraging flock of Red-cockaded Woodpeckers, hearing
their constant short, clear notes coming from pinecone clump areas high in
the trees or from an exposed tree trunk or branch and watching them glide
from tree to tree past me on all sides to forage in other trees. Because of the
open areas where they are found and their noisiness, once I was near a foraging flock I was likely to know it and to find them. I always had the feeling that
they were all keeping up a constant conversation as they fed, chatting about
the day, the yummy food, each other.

When I moved to Texas, I was delighted to realize that some of the East
Texas piney woods are basically clones of the Red-cockaded Woodpecker
habitat that I loved in North Carolina. Some of these areas include the
W. G. Jones State Forest and the Angelina National Forest. The latter was
the place where many of the competing teams in the Great Texas Birding
Classic, including the team that I joined for a couple of years, stood at dawn

each spring in a "secret" spot waiting for the first glimmers of light and the first morning calls of the Red-cockaded Woodpecker to begin our frenetic day. I prefer to encounter them by lazily wandering around the sunny, grassy openings and under the pine trees, listening for and to their sounds, and being surrounded by the calling woodpeckers.

## Red-crowned Parrot (*Amazona viridigenalis*)
Global Population Estimate: 5,000, all in northeastern Mexico and Texas

Red-crowned Parrots are popular cage birds, but their native habitat is in northeastern Mexico. In many locales in the United States where only one or two Red-crowned Parrots can be found in a park or neighborhood, it can safely be assumed that they are probably escaped cage birds. Some of the larger US populations of Red-crowned Parrots, however, such as those that are found in the Lower Rio Grande Valley of Texas and in Florida, Hawaii, and southern California, are "countable" as wild birds, because they are long-established populations, in some cases as a range extension of the northeastern Mexican population.

The twelve-inch long Red-crowned Parrot has a typical parrot shape, with a typical green body and a pale, hooked beak. Its distinguishing features are its red crown above the beak and on the top of the head, a red area on its wings, and a broad yellow tip on its tail. It also has a violet-blue streak behind its eye.

The normal habitat of Red-crowned Parrots is tropical deciduous wooded areas and Tamaulipan scrub in ravines and riparian areas having larger trees. It has adapted well to wooded southern US cities. It nests in hollow trees, sometimes in old woodpecker holes. Red-crowned Parrots are typically seen flying in pairs, their wings flapping quickly in rapid, shallow beats. The food of these parrots includes fruit, seeds, nuts, leaves and buds, and insects.

Before the cage-bird trade became well established, there presumably was less of a threat from bird trappers; however, as with most parrots, the Red-crowned Parrot has always been tempting to trappers. It is believed that most of the colonies of Red-crowned Parrots in the United States have their origin in escaped cage birds, and to the extent that is true, the population of Red-crowned Parrots in the United States has increased substantially since the late eighteenth century.

While in some US areas the numbers of Red-crowned Parrots do not appear to be decreasing and may actually be increasing, there is an overall population decline. In Mexico the Red-crowned Parrot is listed as endangered due to habitat loss and the fact that wild-caught birds, often taken from nests, are still sold as pets by poachers. Poachers also cut down the nest trees of Red-crowned Parrots to get the nestlings, which is harmful both to the birds and to the habitat. Between 1970 and 1982, nearly thirty-five thousand Red-crowned Parrots were exported from Mexico. Additional Mexican habitat has been lost due to government-funded forest clearing. Less than 17 percent of the original forest vegetation used by Red-crowned Parrots in Mexico remains today.

Harvesting wild Red-crowned Parrots is illegal but ongoing. It is important to continue to enforce the ban against importation and smuggling of wild-caught Red-crowned Parrots and, in areas where the birds are caught, to prosecute the offenders who take the parrot chicks and those who fell the nesting trees. Those who are involved in the pet industry, including consumers, need to be sure that wild-caught Red-crowned Parrots are not bought or sold. Anyone who knows or believes that the origin of a Red-crowned Parrot is suspicious should notify the US Fish and Wildlife Service Office of Law Enforcement (http://www.fws.gov/le/) or US Customs.

Because many Red-crowned Parrots are taken from nests in large trees in pastures, it is thought that making agreements with landowners could rapidly reduce removal of parrots from their nests. Exclusion of snakes from parrot nests could also reduce nestling mortality.

A Personal Note: Not too long after I moved to Texas in 2000, I went on a number of field trips in the Lower Rio Grande Valley with the sole purpose of seeing Red-crowned Parrots, but I did not have any luck finding them. Unwilling to accept the status quo, I told everyone I met that I wanted to see a Red-crowned Parrot, and one kind soul told me that Red-crowned Parrots were to be found in Weslaco along a commonly jogged route. The next morning, I stationed myself in a sleeping Weslaco neighborhood and walked up and down the streets a bit so I would not appear to be lurking (which I was). Not too long after it was fully light out, a few large, squawking parrots barreled into a large tree, and I raced over there. I found many more Red-crowned Parrots that appeared to have spent the night in the tree, mostly hidden among the large leaves.

Other years, I have seen Red-crowned Parrots in other Texas locales:

flying swiftly over the Frontera Audubon site in Weslaco, climbing around in the trees in Pharr, and sitting on a power line in Brownsville.

## Black-capped Vireo *(Vireo atricapilla)*
Global Population Estimate: 8,000, all in North America

Black-capped Vireos are the smallest vireo (four and a half inches) in the United States. The bird has an olive-green back, white throat and underparts, and yellowish flanks and wing-bars. The male's black "cap" (the female's cap is dark gray) actually extends from the top of its head down the nape and on to the cheeks. Broken white spectacles on the vireos' mostly black heads are a very distinctive field mark. In the United States they are found only in Texas and Oklahoma in areas with a mix of shrubs and open grassland and considerable woody foliage under ten feet tall. In Texas, where the highest concentrations of these vireos are, they are found primarily in the central Edwards Plateau and eastern Trans-Pecos regions. In Oklahoma, they are found in three counties, primarily in the area of the Wichita Mountains. They are in the United States only in the summer and spend the rest of the year in western Mexico. Some also breed in Mexico.

Each year Black-capped Vireos return to the same two-to-four-acre breeding territory, where the male sings nearly nonstop. The areas where they are found do not have tall trees but scattered scrubby, woody plants such as shinnery oak, sumac, and Texas persimmon, often growing in poor soils or where there has been a fire or other major disturbance of the land. The vireo's cup-shaped nests are built in the fork of a low branch, often a shinnery oak or sumac. Their diet is primarily limited to insects such as the larvae of moths and butterflies, flies, leafhoppers, and the like, as well as spiders.

Black-capped Vireos were once widespread from Mexico to Kansas and were found in a much larger area than they are today. It is believed that their range extended from south-central Kansas through central Oklahoma and Texas and then south and west to central Coahuila in Mexico and in Big Bend National Park in Texas. They are largely extirpated from their former breeding range in the United States. Some of their breeding areas in Oklahoma and all of them in Kansas have disappeared.

The number of Black-capped Vireos was as low as about 256 to 525 breeding pairs in 1987, when the species was listed as endangered. Since that time, there have been increases in the number of birds, but the total number still

remains very low. Most of the US population (about 75 percent) is found in only four areas: Fort Hood Military Reservation (Texas), Kerr Wildlife Management Area (Texas), Fort Sill (Oklahoma), and the Wichita Mountains National Wildlife Refuge (Oklahoma). In 2013 a report by the Southwest Region Ecological Services office of the US Fish and Wildlife Service found that the Black-capped Vireos may no longer be in imminent danger of extinction and that reclassification from endangered to threatened may be warranted.

Black-capped Vireos are favorite targets of Brown-headed Cowbirds, which lay their eggs in the vireo nests, often causing the vireos to abandon the nest. If they do not abandon their nests, they raise the young cowbird and thus fewer of their own young survive. Other predators, such as cats, raccoons, and skunks, are also a threat to Black-capped Vireos.

People who own property in areas of Texas and Oklahoma where Black-capped Vireos are still found can manage their property for the vireos. Information on what can be done to manage a site for Black-capped Vireos is found at the website of the Texas Parks and Wildlife Department (TPWD, at http://www.tpwd.state.tx.us/). Recommended measures include prescribed burning every five to seven years during the cool season to keep Ashe juniper (cedar) invasion in check and allow regrowth of broad-leaved shrubs, selective brush management by moderate thinning, grazing and browsing management so that the thick, woody growth needed for nest concealment is not eaten, and reducing cowbird numbers by removing grazing animals from vireo nesting areas and/or trapping or shooting cowbirds, which in Texas requires certification by the Texas Parks and Wildlife Department.

A Personal Note: My first views of Black-capped Vireos were in Texas after I moved to the state in 2000. Each spring they arrived from their Mexican winter habitats and with luck could be seen (or at least heard) singing in their limited Central Texas range. One of my favorite areas, part of the Balcones Canyonlands National Wildlife Refuge northwest of Austin, Texas, is closed during part of the breeding and nesting season but is open the rest of the time and is one of the best places near a major urban center to see a Black-capped Vireo. I needed to get there either in very early spring or late spring (when the birds were less likely to be heard singing) or after the site reopened. Usually there would also be a singing White-eyed Vireo and a very noisy Yellow-breasted Chat making it difficult to hear the singing Black-capped Vireos, which usually were nearly impossible to see. There are also guided tours at this refuge, including during the annual Balcones Songbird Festival.

My best views of a Black-capped Vireo were in the Wichita Mountains of Oklahoma, not too far north of Texas. In 2010, when I was taking trips from Texas to explore Oklahoma birding sites, I attended a meeting of the

Oklahoma Ornithological Society held near the mountains. On a Sunday morning field trip we went to the Wichita Mountains and had very close views of at least one Black-capped Vireo, but the trip moved on to look for other birds and I was not able to get a very good picture. So, I came back the next morning and hung around in the same general area until I finally found a singing vireo. I carefully stalked it as it sang in a low sapling or brush, then disappeared awhile, usually still singing, while it presumably had breakfast, and then started singing from another area. While some of its reappearances were distant and not photographable, some of them were very close to me, and just a couple of times I was able to get a picture or two between or over the leaves.

FLORIDA SCRUB-JAY AND ISLAND SCRUB-JAY

Florida Scrub-Jays and Island Scrub-Jays are both noisy gray-and-blue jays, closely related to, and once classified with, the much more common Western Scrub-Jay.

## *Florida Scrub-Jay (Aphelocoma coerulescens)*
Global Population Estimate: 6,500, all in Florida

The Florida Scrub-Jay is the only bird species that is endemic to Florida. Florida Scrub-Jays are one to two inches longer than an American Robin and about the size and shape of the common Blue Jay. Although most jay species found in the United States are primarily blue, the patterning and type of blue color differ among jay species. The Florida Scrub-Jay, unlike the Blue Jay, does not have a crest. The blue color in the Florida Scrub-Jay is a pale blue, especially on the head and neck. The jay's wings are a plain blue, without the black barring and white wing-patches of the darker, blue-purple Blue Jay. The Florida Scrub-Jay has a whitish forehead and area over the eye and a pale gray belly and upper back.

Like other jays, Florida Scrub-Jays eat a wide variety of food, with their favorites being insects and acorns, the latter of which are often hidden in the ground by the jays, especially in sandy areas, for storage. They often become quite tame, as is the case with some other jay species, if they become used to the presence of humans.

Florida Scrub-Jays are found only in central (peninsular) Florida, generally being found in scrub and scrubby flatwoods habitats where there are sufficient numbers of scrub oaks to provide a large enough acorn supply for the wintertime, as well as places to find cover from predators and nesting sites (March to June). The vegetation at the preferred sites grows primarily on very well drained sandy soils on the coast and other sandy dunes and ancient shorelines on the peninsula. The habitat vegetation includes various evergreen oaks, preferably not more than a couple of yards high, and sparse

ground cover consisting of such plants as palmettos. The areas in which Florida Scrub-Jays are found are often subject to fires, which help maintain the Florida scrub ecosystem in which these jays thrive.

Florida Scrub-Jays are more "family" oriented than many bird species. They live in groups as small as two birds and as large as extended families of eight adults and one to four young birds. Once baby Florida Scrub-Jays leave the nest, they stay in the nesting area with their parents and serve as "helpers." Eventually, after one to seven years, young males find their own territory and a mate and start a new family group.

Florida Scrub-Jays once were found in thirty-nine of the forty counties in peninsular Florida; however, habitat destruction has resulted in a substantial decrease in their numbers. In some areas, their population has decreased by 80 percent or more, and, in others, they have been completely extirpated. It is estimated that the population has decreased by 25 to 50 percent in a recent ten-year period.

The Archbold Biological Station at Venus in south-central Florida is an ecological research institute and preserve of more than five thousand acres that supports about one hundred Florida Scrub-Jay families (http://www.archbold-station.org/). The primary focus of the Archbold Biological Station is on the organisms and environments of Lake Wales Ridge and adjacent central Florida, including avian population biology, demography, and regulation, as well as endangered species (such as the Florida Scrub-Jay) management and conservation.

The US Fish and Wildlife Service is working with the Merritt Island National Wildlife Refuge, the Cape Canaveral Air Force Station, and the Ocala National Forest, as well as with the state and local governments and private landowners, to manage the land appropriately for the Florida Scrub-Jay. The annual Florida Scrub-Jay Festival is a collaborative effort of multiple conservation organizations and helps inform the public about the Florida Scrub-Jay.

Private landowners in peninsular Florida who are interested in having habitat improvements that are of benefit to the Florida Scrub-Jay can contact the US Fish and Wildlife Service as provided under the Endangered Species Act.

A Personal Note: Like the Whooping Crane, the Florida Scrub-Jay was a species I thought I might never have a chance to see. While North Carolina does not seem particularly far from Florida, I somehow was never able to get down there to the jays' area, but instead my few trips to Florida were for meetings in Miami Beach without time to head north from there to find jays. Once in the mid-1980s, when my husband had a meeting in Orlando, I decided to accompany him on the drive and commandeer the car during his meeting so I could go birdwatching, and then we could look for Florida Scrub-Jays after the meeting as we drove north. When we drove into the habitat north of Orlando where the jays were supposed to be, we easily found a small flock of them gliding between the scrubby trees.

My second (of two) Florida Scrub-Jay sightings was in 2008, when they were not quite so easy to see. My North Carolina friend Lena and I finally located two of them after taking an extended hike through the entire Juno Dunes Natural Area north of Fort Lauderdale. When we returned to our rental car back at the parking lot after our unsuccessful walk, two Florida Scrub-Jays were hanging out on the scrubby vegetation surrounding the lot, as if they were waiting around for us to find them.

## Island Scrub-Jay *(Aphelocoma insularis)*
Global Population Estimate: 2,500 to 9,000, all on one California island

Island Scrub-Jays, found only on Santa Cruz Island, a large island off the coast of California, were once grouped in the same species classification as Western Scrub-Jays, found in the western United States. Genetic studies, as well as differences in coloration between the Island Scrub-Jays and Western Scrub-Jays, led to their being classified as separate species. All three species of Scrub-Jays have blue upper parts and white underparts and a blue-gray breast band, but the Island Scrub-Jay is a brighter blue and is larger. Because there are no other jays on Santa Cruz Island, there is no problem in identifying any jays seen there.

Coastal live oak woodlands and chaparral make up the habitat where Island Scrub-Jays breed. They are monogamous and very territorial, holding the same territory year after year. Nonbreeding birds roam the island without being in defined territories.

After the settlement of California in the early history of the United States, sheep and goats were introduced on some of the Channel Islands, which caused destruction and degradation of the habitat required by the Island Scrub-Jays. Santa Cruz Island is part of Channel Islands National Park, and there is an effort being made to control the introduced animals and improve the islands' habitat.

Although the published estimate of the number of Island Scrub-Jays in most references is nine thousand, a detailed study (2012) coauthored by Smithsonian Conservation Biology Institute scientists found a much lower number of twenty-five hundred.

Channel Islands National Park can be supported by donation and by volunteering to staff the information desk, serve as a volunteer naturalist, help maintain trails, or restore vegetation, or by participating in other activities (http://www.nps.gov/chis/supportyourpark/index.htm).

A Personal Note: I have been to Santa Cruz Island only once, in September 2008, although I tried to go there earlier than that but was deterred by heavy rain. When I finally was able to go, I got on the boat along with hordes of other people, most of whom were going to other places. Along with other hard-core birders I was dropped off at Prisoners Harbor on the island. While most of the birders stayed near the dock area at first, I hiked out the road toward what I thought were the distant sounds of jays. I found nothing and wandered back. Another group saw some jays that I did not see. Although I was somewhat worried, I figured that it was an island after all and we were there all day, so eventually I would see one. After a couple of hours of random wandering, I heard happy birder shouts and went back to the dock, where two Island Scrub-Jays were very visibly and noisily hopping from tree to tree to ground, ignoring the excited bird-people.

## *California Gnatcatcher (Polioptila californica)*

Global Population Estimate: 77,000, all in California and Mexico

California Gnatcatchers are small, slim gray birds that are similar in size and shape to the widespread and much more common Blue-gray Gnatcatchers, being about four and a half inches long. They have long, thin, white-edged tails and a thin white eye-ring, and the breeding males have a black cap. They are found year round in low, arid scrub areas along the southern California coast and in Baja California, Mexico.

Little is known about the population of California Gnatcatchers prior to the 1990s. Many areas once inhabited by California Gnatcatchers were no longer usable by the gnatcatchers after the expansion of communities reduced and fragmented the gnatcatcher habitat. In 1996, after California Gnatcatchers were listed as being threatened, it was estimated that there were about three thousand pairs in the United States, not including the birds in areas where their habitat was being destroyed completely, in addition to a similar number of California Gnatcatchers in Baja California, in Mexico.

The US Fish and Wildlife Service has designated thirteen critical habitats of California Gnatcatchers under the Endangered Species Act, and most of them are on privately owned land. Because California Gnatcatchers are regularly parasitized by cowbirds, management in these habitats regularly

includes trapping of cowbirds. California Gnatcatchers are also very susceptible to cold, wet winters and can have their numbers greatly reduced when
such a winter occurs.

In addition, California has enacted legislation to protect coastal scrub
areas where the California Gnatcatcher is found, while still allowing limited
development in these areas. California's Natural Community Conservation
Planning (NCCP) program (https://www.dfg.ca.gov/habcon/nccp/) is a partnership between the state and various public and private groups using an
ecosystem approach to protect plants and animals and their habitats while
allowing economic activity that is compatible with the protection efforts.

Conservation of the California Gnatcatcher is specifically stated as being the "dominant factor" in determining how this program is implemented.

In June 2014, however, various property rights groups and developers took actions to try to remove the California Gnatcatcher from being listed under the Endangered Species Act, stating that the California Gnatcatcher subspecies in California is identical to the more abundant subspecies of California Gnatcatcher found in Baja California and is not a valid subspecies. At the end of December 2014, the FWS announced that it would review the status of the California Gnatcatcher to determine if it is qualified as a threatened species under the Endangered Species Act.

A Personal Note: I have seen California Gnatcatchers only once. In early September 2008, after unsuccessfully checking a couple of sites in southern California, I drove to one of the developments, Ocean Trails, which was touted as being made especially to take California Gnatcatchers into account. At first there was no activity in the bushes along the trail, but then I saw first one and then three tiny, dark, long-tailed gnatcatchers and heard their distinctive mewing sound in the low, scrubby bushes.

## Bicknell's Thrush (Catharus bicknelli)
Global Population Estimate: 40,000, all in North America

The Bicknell's Thrush is related to the American Robin but is about three inches shorter. It has olive-brown upper parts and gray-white underparts, with dark spots on the upper breast.

The Bicknell's Thrush breeds only in northeastern North America, with its breeding area being limited to high-elevation forests in the northeastern United States (mountains of Maine, New Hampshire, Vermont, and New York) and mountain highlands of northeastern Canada (northern Gulf of St. Lawrence and Nova Scotia and southward). The forests where it is found are usually along ridgelines and have a dense understory with thick bushes and thickets. Such habitats are often found in areas that have been disturbed so that new stands of balsam fir are growing amid larger dead fir and spruce trees. Typically, the nests are in thick stands of young spruce or balsam fir trees. In fall, Bicknell's Thrushes migrate to the Caribbean, where they winter. Bicknell's Thrushes eat insects and other invertebrates that they find as they hop on the forest floor.

Bicknell's Thrushes were discovered in the Catskills in 1881 and were at

first classified as being the same species as Gray-cheeked Thrush, which is not endangered and nests across all of northern Canada. It is unlikely that there was much in the way of early data on the Bicknell's Thrush population size because of the species' remote nesting area and retiring ways. It is likely that the population was much larger when there were substantially more uninhabited forested areas in which it could breed.

Because the Bicknell's Thrush breeds in such a small area of North America and winters in such a small area of the Caribbean (primarily Hispaniola) and because there are so few of them, there is concern that environmental changes caused by shifts in the climate or human activity, such as logging, pollution (air and water), acid rain, recreational development, and the building of structures such as telecommunications equipment and wind power facilities, are detrimental to Bicknell's Thrush population numbers.

The International Bicknell's Thrush Conservation Group (IBTCG, at http://www.bicknellsthrush.org/) was formed in 2007 by a coalition of scientists, natural resource managers, and conservation planners. The goal of the group was to implement a conservation action plan, and it was released

in 2010. The overall mission of the group is to develop a broad-based, sci-
entifically sound approach to conserving the Bicknell's Thrush, using monitoring and on-the-ground management actions. Specifically, the goals are
to increase the global population of Bicknell's Thrush by 25 percent over
fifty years, ensure that there is no further net decrease in distribution of the
thrush in both its breeding and its wintering grounds, implement and sustain
a range-wide breeding season monitoring program, and implement direct
conservation and research actions to address identified threats to Bicknell's
Thrushes, leading to improved protection, management, and restoration of
breeding and wintering habitats. In 2012 Grupo Jaragua, a partner of IBTCG,
reported a serious ongoing loss of broadleaf forest in a Bicknell's Thrush wintering area in the Dominican Republic.

There are a number of ways to help Bicknell's Thrush populations. Learn
about, volunteer for, and support Mountain Birdwatch (http://www
.vtecostudies.org/MBW/), which is an offshoot of the Vermont Forest Bird
Monitoring Program and is supported by the US Fish and Wildlife Service,
the Stone House Farm Fund of the Upper Valley Community Foundation, several conservation groups and government agencies, and individual
donors. It was launched in 2000 with the goal of establishing a long-term
monitoring program for Bicknell's Thrush, as well as other montane forest
birds. Trained volunteers conduct dawn surveys in the awe-inspiring forests
where the Bicknell's Thrush breeds. A typical time commitment for such a
volunteer is one to two mornings each June.

One can also assist in monitoring Bicknell's Thrushes in New Brunswick
and Nova Scotia (http://www.bsc-eoc.org/volunteer/achelp/index.jsp) and in
Quebec (http://www.bsc-eoc.org/volunteer/achelp/index.jsp).

A Personal Note: Having never lived in the northeastern United States, I did
not see a Bicknell's Thrush until a few years ago. I had tried to see one years
earlier during a July nonbirding trip to Rhode Island, and while driving up
and down various roads I saw only Swainson's Thrushes. In 2008, however, I
scheduled a guide (Derek Lovitch) to help me find Bicknell's Thrush habitat
and, I hoped, a Bicknell's Thrush. Rather than take the easy route (drive to a
good area, play a recording of a Bicknell's Thrush, and hope for a responsive
thrush), we chose to hike, with full backpacks, up a downhill ski slope on
Saddleback Mountain in Maine. As we hiked up on an early July afternoon,
Lovitch regaled me with tales of other hikers he'd taken up to look for the
thrushes. After we wandered around a bit, set up our tents, and had supper,

we heard a Bicknell's Thrush calling very near our tents. One of them blasted across the opening formed by the ski trail and dove into a very dense stand of short spruce trees. That was our only sighting of Bicknell's Thrush, and all we saw after that were Swainson's Thrushes. We did hear the Bicknell's Thrushes, however, which in itself is as "countable" as an actual sighting under the American Birding Association rules. The next morning I awoke at 4:24 to the song of a Bicknell's Thrush.

## McKay's Bunting *(Plectrophenax hyperboreus)*
Global Population Estimate: 6,000, only on western Alaskan islands

One must make a huge effort to see a McKay's Bunting because they usually breed only on two small islands (Hall and St. Matthew) in the Bering Sea off Alaska, sometimes wandering to St. Lawrence and St. Paul Islands, which are easier for birders to visit than where the species normally breeds. Because of the remote areas in which McKay's Buntings are usually found, population estimates can be based only on the known density in the areas where they have been counted, and as a result the estimates of their population numbers vary widely.

McKay's Buntings, like Snow Buntings, are mostly all white but typically have a smaller black area on the wing-tip and a much smaller black area on the tip of the tail than do the Snow Buntings. These mostly white bunting species both have a low warbling call and live in the same habitat. It is believed that McKay's Buntings may have evolved from Snow Buntings, and, unfortunately for those wishing to be definitive in their identification of these birds, the two species sometimes hybridize in the small geographic area where they are found together.

The preferred habitat of McKay's Buntings is open rocky ground, beaches, and shores of tundra pools. McKay's Buntings nest on shingle beaches in crevices and hollow drift logs on the islands where they breed, and they winter on shingle beaches, in marshes, and in agricultural fields along the western and southeastern coast of Alaska, from Kotzebue south. They eat weed and grass seeds, as well as buds and insects if available.

It is unlikely that early naturalists knew about McKay's Buntings, and therefore little is known about their population size then. Because of their limited and remote habitat, it may be that the population of McKay's Buntings has always been relatively small. Although the known population of

McKay's Buntings is and may always have been very small, it is regularly included in species lists of birds that are "at risk." It is unclear whether such a small population can be maintained. The main concern at this point is that there needs to be more information on this species. It is also clear that, like other island-nesting species, McKay's Bunting would be extremely threatened if rats or other non-native animals were to become established on their nesting islands. It would not take much of a change in human behavior on or near these islands to tip the population from being apparently low in numbers to being on the verge of extinction.

It is important to determine the current population density and population trends for the McKay's Bunting to see if the population is stable or threatened. Because of the small population size it is evident that any threat to the nesting islands could threaten the entire population. Studies on the breeding grounds are necessary to document breeding behavior and nesting success.

Although most of us living far south of where McKay's Buntings can be found cannot help the McKay's Bunting directly, we can support and

participate in organizations such as Partners in Flight (http://www.part-nersinflight.org/). Partners in Flight describes what needs to be done for McKay's Bunting in the PIF Species Assessment (http://www.pwrc.usgs.gov/pif/watchlistneeds/MCBU.htm).

A Personal Note: I had been to both St. Paul Island and Gambell on St. Law-rence Island more than once without seeing a McKay's Bunting. In 2008, while I was on St. Paul Island in May, someone reported a McKay's Bun-ting there. Sadly, once we raced out and then found and photographed the very white bird after working our way over rocks and grass tussocks, it was decided that evening by the experts who reviewed the photographs that it was a hybrid McKay's-Snow Bunting. This news meant that the photographed bird was not a member of a full species and thus not a countable McKay's Bunting. For a while, in my desperation to get a new bird species, and espe-cially that new bird, I was going to count it as three-quarters of a McKay's Bunting (it was thought to look more like a McKay's Bunting than the other parent, a Snow Bunting). Of course, that was not really acceptable.

Thankfully, on June 3 at Gambell, we saw a "real" McKay's Bunting, some-what whiter than the one on St. Paul, and no one questioned its parentage. It turns out that the experts, when they again reviewed photos of the St. Paul Island bird, changed their minds and decided that it really was a McKay's Bun-ting. Therefore, that year I saw two of them, and that is all I have ever seen.

### Colima Warbler (Oreothlypis crissalis)
Global Population Estimate: 25,000, all in Mexico and Big Bend National Park in Texas

Colima Warblers are large as warblers go (about five and three-quarters inches long) and relatively drab as warblers go as well. They are mostly brownish green (back) and gray (underneath and face) with a white eye-ring. The most noteworthy coloration of a Colima Warbler is a bright yellow rump and vent and a small splash of rusty color on the top of its head, which may be difficult to see.

The only place that Colima Warblers breed in the United States is the Chisos Mountains of Big Bend National Park in southwestern Texas. They also breed in the Sierra Madre Occidental of northwestern Mexico. They winter on the Pacific slope of southwestern Mexico. In this narrow area, the particular habitat where the Colima Warbler prefers to breed is at relatively high elevations (1,500 to 3,600 meters, or roughly 5,000 to 12,000 feet) in dry,

shrubby chaparral of oaks and pines with a grassy ground cover. They nest on the ground in the leaf litter or grass or in cavities on the hillsides, generally where there is vegetation overhead. Like many warblers, Colima Warblers eat insects.

Because of their extremely limited range in the United States, and possibly due to range expansion in the 1900s, it was not known until 1928 that Colima Warblers bred north of the Mexican border, and therefore the status of the Colima Warbler in the early days of the United States is unknown.

Although most publications on the status of birds in the Americas mention the Colima Warbler and concern about its low population numbers, it is not clear if there is an imminent problem. It is important to continue to preserve the Chisos Mountains in Big Bend National Park, and therefore governmental and private threats to this habitat should be monitored in this time of cutbacks in federal spending.

CIPAMEX (http://www.cipamex.org/) is the National Audubon Society BirdLife International Partner in Mexico. It has an Important Bird Area program and is working to protect Colima Warbler habitat (and habitats of other species). This partnership and these organizations should be supported. There are also ways to help Big Bend National Park itself (http://www.nps .gov/bibe/index.htm), including supporting the Big Bend Natural History Association and Friends of Big Bend National Park, as well as by volunteering at the park itself.

A Personal Note: The day that I first saw a Colima Warbler I thought I was going to die. As a new Texas resident in 2000, I had signed up for a Big Bend National Park field trip at an April meeting of the Texas Ornithological Society. Our group hiked the Pinnacles Trail more than four miles up, up, and more up, nearly nonstop it seemed to me, until we reached an area with deciduous saplings along the trail and a view through the saplings all the way down to the Chisos Basin parking lot from which we'd started the hike. Finally, we halted to listen for, and then to hear, the distinctive song of a Colima Warbler. Without too much effort—which was good because I could not have made more effort—we located one of the singers, nearly down at eye level. I was sure it would be my one and only view of a Colima Warbler. When someone on the hike mentioned that there would be even more of them up there later in summer, I was sure that I was not going to hike the trail in summer considering how overheated I was from the climb in April. What I did not know was that in late summer, the Chisos Mountains usually cool down to the fifties at night and are very pleasant during the day due to the monsoons, even though the lower elevations continue to be sweltering.

As time passed, the agony of that hike receded from my memory, and I considered doing it again. This time, however, I had other goals (see the Flammulated Owl account above), and I decided to join a group backpacking hike in late July. Others in the group were clearly in better shape than I was, but I managed to make it to the top of the four-and-a-half mile climb with a heavy backpack (binoculars, camera and telephoto lens, bird book, warm clothing for the fifty-degree nights, food, water, sleeping bag). We did hear and see quite a few Colima Warblers on the trek, and the agony of the backpack-laden climb was forgotten (or at least repressed).

My love of the Chisos Mountains and my annual listing mentality made me go back nearly every year while I lived in Texas for at least one and sometimes two hikes up the mountain, usually at least one of which was with a heavy backpack. I even volunteered a couple of times to take someone else up the mountain for their first view of a Colima Warbler, but usually I, the leader, became a follower as I struggled once again to climb the mountain behind the usually younger, spryer birders. It was always worth the effort, however, to see a bird that is so rare in the United States at the only place that it can be found north of the border.

## Kirtland's Warbler *(Setophaga kirtlandii)*
Global Population Estimate: 4,500, all in North America and the Bahamas

Kirtland's Warblers are large for a warbler, about five and three-quarters inches long. They have a dark, slate-gray back and face, white crescents below and above their eyes, a yellow throat, breast, and belly, two thin white wing-bars, and black stripes down their sides. As is true with a few other warbler species, Kirtland's Warblers constantly wag their tails.

The only place that Kirtland's Warblers have been found to nest is in jack pines, and for that reason they are also informally known as jack pine warblers. Until the mid-1990s, they were thought to nest only in the jack pines of Michigan's Lower Peninsula. Since then, they have also been found nesting in the Upper Peninsula, as well as in Wisconsin and Canada. They winter in the Bahamas.

As with my experience in seeking information on the other warblers found only in a few locales, I was unable to find evidence that the Kirtland's Warbler was generally known by early US settlers. Although early naturalists such as Audubon did come to Michigan in the 1830s, the Kirtland's Warbler was discovered near Cleveland, Ohio, only in May 1851, presumably during migration.

The small size of the population of Kirtland's Warblers is the problem. When the first census was conducted in 1971, there were only 201 singing males, all of them in Michigan. The good news is that current management practices are helping to increase the population. The first birds singing

outside that area were found in 2007. At the end of the breeding season in 2011, results of the annual census were released. They indicated that in central Michigan there were 1,170 singing males counted, with an additional 35 birds in the Upper Peninsula. In Wisconsin there were 21 singing birds, while Ontario had 2. Increased efforts to manage the jack pine barrens where the birds have been documented to be singing seemed to be working in Wisconsin, where there were eleven recorded nesting attempts in 2011, with four nests known to be successful. The Wisconsin Department of Natural Resources continues to monitor Kirtland's Warbler singing and nesting activities in seven counties. In 2012 there were 2,090 singing males in Michigan and Wisconsin (http://www.fws.gov/midwest/news/602.html). These numbers, while encouraging, still indicate that Kirtland's Warbler populations are small enough that they need to be monitored on their breeding grounds, as well as on their wintering grounds in the Bahamas, to determine habitat requirements and actual wintering locations. It is important that wintering areas not be cleared for resorts or other purposes; however, the inhospitable environment away from the coast in the Bahamas means that attempts to cultivate and develop these areas are often not successful and the land is then abandoned and left to revert to forest, which is favorable for these warblers.

The Michigan Department of Natural Resources (http://www.michigan.gov/dnr) developed the Kirtland's Warbler Recovery Plan in 1976 and updated it in 1985, and it provides state and federal agency personnel with a structured guide for managing Kirtland's Warblers. The objectives include developing and maintaining about 40,000 acres of suitable nesting habitat for Kirtland's Warblers through a careful rotation of timber harvesting on 140,000 acres of jack pine stands, protecting the Kirtland's Warbler on its wintering grounds and along its migration route, reducing key factors affecting reproduction and survival of Kirtland's Warblers, monitoring their breeding populations, and developing and implementing emergency measures to prevent extinction of Kirtland's Warblers.

The Michigan Department of Natural Resources invites people to become partners in helping the Kirtland's Warbler by supporting its own efforts, and it provides a list of how you can help. The list includes commonsense rules about not doing things to disturb the Kirtland's Warbler (not trespassing, camping only in designated areas, not using recordings) and donating to the Nongame Fish and Wildlife Fund.

In the late 1990s agencies in the United States and the Bahamas formed a partnership to identify and protect the winter habitats of Kirtland's Warblers and other warblers. Participants in this partnership include, among others, the nonprofit Bahamian National Trust (http://www.bnt.bs/How-to-Get-Involved), membership in which supports addressing environmental concerns and environmental education in the Bahamas; The Nature Conservancy (http://www.nature.org/), which has worked to protect land in the Bahamas; and the Bahamas Ministry of Agriculture and Fisheries. It is important to support the Bahamian government's efforts to protect wintering areas for these warblers.

A Personal Note: Although I grew up as an avid birdwatcher in central Wisconsin on land covered with jack pines (but not Kirtland's Warblers), only a state away from where almost all of the world's population of Kirtland's Warblers nests, I did not try to see one until the late 1990s. My husband and I had gone from North Carolina to Wisconsin to visit my father. Midway through our Wisconsin visit, we scheduled a birding trip to Mio, Michigan, one of the two main areas in Michigan where the US Fish and Wildlife Service, the US Forest Service, and the Michigan Audubon Society lead trips to look for Kirtland's Warblers each spring. When we reached the Wisconsin shore of Lake Michigan and found the ferry dock, we also found that there was a serious problem with the ferry, which resulted in a very long trip and an early morning arrival in Michigan. We drove off the ferry toward Mio, and although there were deer everywhere along and on the road and bounding across it, we made it and arrived at our motel an hour or so before we were due to meet the ranger. I slept an hour and then went to the ranger station, leaving my nonbirding spouse sleeping soundly.

I did see Kirtland's Warblers that morning, and did hear them, serenading the group from the jack pines extending for miles in the carefully managed habitat. I even stayed awake long enough to get back to the motel, wake up my husband, and take him back with me so he could see the warblers too (whether he wanted to or not).

My second trip to see Kirtland's Warblers was far less eventful—a simple flight and a drive to Mio, followed by sight and sound, but not as close a view as before. It was still a wonderful thing to see such a rare bird.

## *Cerulean Warbler (Setophaga cerulea)*
Global Population Estimate: 560,000, all in the Americas

The word *cerulean* means "resembling the blue of the sky," and it perfectly describes the male Cerulean Warbler. The back and top of the male's head is a bright sky blue, and he is white underneath with a narrow breast band and dark streaks on the side of his breast. The adult female is a more subdued, grayish, blue-green color instead of sky blue. Both of them have white wingbars and white tail-spots that can be seen when they fly.

Cerulean Warblers breed primarily in the eastern United States in

deciduous forests having mature broad-leaved trees and generally some areas
that open to the sun. They also can occur in other wooded areas. They currently breed from Arkansas to the northeastern and central United States up
to southern Ontario, and they winter in the mountain forests of northern
South America.

Cerulean Warblers were once one of the most abundant breeding warblers
in the valleys of the Ohio and Mississippi Rivers, but destruction of large portions of their breeding habitat in the 1900s caused the numbers to plummet.
Their traditional primary concentration was in the old-growth bottomlands
of the Mississippi River, but these forests are gone.

The Bird Conservation Alliance calls the Cerulean Warbler "North America's fastest declining migratory songbird," a decline due primarily to habitat
destruction. It is estimated from Breeding Bird Surveys that between 1966
and 2001 the population of Cerulean Warblers decreased about 4.5 percent
per year. While the population of Cerulean Warblers appears to be large, the
magnitude of the threats to Cerulean Warblers is also very great.

Efforts are being made by a number of groups to acquire habitat, preferably large tracts that do not require management. The Cerulean Warbler Atlas Project of the Cornell Lab of Ornithology is designed to uncover more information about the status and needs of Cerulean Warblers. Much of the suitable habitat for Cerulean Warblers is in national wildlife refuges, and it is therefore critical that funding for the refuges be maintained.

The Bird Conservation Alliance (facilitated by the American Bird Conservancy, http://www.abcbirds.org/birdconservationalliance) works to raise money to purchase land containing important warbler wintering habitat in Colombia, to train forest guards and guides, and encourage ecotourism businesses. Donations are encouraged.

A Personal Note: Nearly every year that I lived in North Carolina I went to the North Carolina mountains to look for warblers, and I was always delighted when I had a brief glimpse of a Cerulean Warbler high in the treetops, especially if it was the bright blue male. One year stands out especially in my mind, however. It was on a field trip that was part of a Carolina Bird Club spring meeting. I'm sure we saw other warblers, but all that I remember was a single huge tree surrounded by other trees stretching into the distance, sunlight filtering through the leaves. In that tree were five male Cerulean Warblers flitting about, chasing each other, showing off their blues like little, winged, glowing jewels.

My other noteworthy memory was of a single Cerulean Warbler on the Texas coast. Again it was spring and I was part of the annual birder migration to the coastal habitats to see migrating birds that, having flown across the Gulf of Mexico, land exhausted in the small coastal trees and brush after their flight. I was walking along a roadway not far from the shore and saw a small bird on the ground sitting on the roadside gravel. I walked over and looked down at the bluest back of a very tired warbler, hopping, picking at insects in the dirt, totally unconcerned about me looming over it, and maybe even unaware of me. It made me realize that not only did Cerulean Warblers face threats at both ends of their migration routes but along the route they also faced exhaustion, starvation, and predators that could pick them off one by one.

## Golden-cheeked Warbler (*Setophaga chrysoparia*)

Global Population Estimate: 21,000, all in Texas and Latin America

The Golden-cheeked Warbler has been called a "smart-looking" black-yellow-and-white warbler. The golden cheek that gives it its name is somewhat similar to the cheeks of a few other warblers that also can have a black throat and sometimes black streaks on their sides (e.g., Black-throated Green Warbler, Hermit Warbler, and Townsend's Warbler), but the Golden-cheeked Warbler also has a single black eye-line, a black (male) or dark olive (female) back, and pure white underparts.

Golden-cheeked Warblers breed in juniper-oak woodlands in Central Texas, primarily in and around Austin and San Antonio. They use the bark of the juniper to make their nests. Because Golden-cheeked Warblers are usually found in well leafed out trees or conifers, they may be difficult to spot; however, they can usually be located by their distinctive but somewhat variable song, which has been described as a buzzy "*zee zeedee-zee.*" Many birders come to Texas with the main goal of finding a Golden-cheeked Warbler, since it breeds only in Central Texas. After spending only about three months in the United States each year, the Golden-cheeked Warbler winters in southern Mexico and Central America.

The food of Golden-cheeked Warblers is limited to insects, which the birds either eat off of branches and leaves or catch from the air like flycatchers.

While I am unaware of any information from early US naturalists, it is known that the breeding range of Golden-cheeked Warblers has decreased dramatically to an area of about sixty-six thousand acres, much of which is part of Fort Hood in Texas.

Efforts to preserve and protect nesting habitat for the Golden-cheeked Warbler are critical. Balcones Canyonlands National Wildlife Refuge, just north of Austin, as well as a number of smaller Texas refuges, have been established in an effort to maintain suitable Golden-cheeked Warbler nesting areas. Brown-headed Cowbirds are frequent in Golden-cheeked Warbler habitats, and, as in the case of Black-capped Vireos, the cowbirds lay their eggs in the nests of warblers, including the Golden-cheeked Warblers. Severe declines in Golden-cheeked Warbler reproductive success have been reported in these areas due to young cowbirds usurping the baby warblers' nest space and the parents' feeding efforts, leading to ongoing efforts to trap cowbirds.

Research indicates that rat snakes, as well as other predators in Texas, can also be detrimental to nesting success of Golden-cheeked Warblers. Therefore, further research on habitat modification to reduce the effects of these predators may be beneficial to Golden-cheeked Warblers.

One way to help Golden-cheeked Warblers is to support Balcones Canyonlands National Wildlife Refuge through donations to the Friends of BCNWR. Donated funds support land acquisition and education and recreation opportunities at the refuge. Volunteers are also needed for such things as maintenance, assistance with special events, outreach, education, fundraising, and public relations (http://www.friendsofbalcones.org/volunteer).

A Personal Note: My first attempt to see a Golden-cheeked Warbler was nearly a total miss. I decided that I would look for them in Meridian State Park in Bosque County in Texas, which was supposed to have many of them in its wooded, juniper-covered hills and which was closer to my home in Fort Worth than were other good areas for Golden-cheeked Warblers. When I got there, very early in the morning and eager to find them, the park gate was still closed, so I amused myself by gazing into the trees that I could see from outside the park. When I got into the park, I took a relatively lengthy walk through the junipers and did not see a single warbler. I did hear what seemed to be a buzzy song, but I had not studied my birdsongs before going in and was unsure whether I was hearing Golden-cheeked Warblers or not. When I got back to my car I got out my tape of bird songs and played the song of the Golden-cheeked Warbler to see if that was what I had been hearing. Not only was it clear that I had been hearing them, but one of the Golden-cheeked Warblers near the parking lot flew toward me and tried to get in my car to find the bird on the tape—definitely a bird that is responsive to taped calls. This is a good place to emphasize that playing bird song recordings, especially for endangered birds, should not be done in areas where the bird species breeds or whenever playing the recording is likely to agitate or disturb these species (see the section on the ABA Code, discussed in the next chapter).

In 2008 I learned in March that Golden-cheeked Warblers were being seen at Warbler Vista, part of Balcones Canyonlands National Wildlife Refuge in Central Texas. I parked in the parking lot, grabbed my camera, and took the trail from there. I hadn't been in the trees more than three minutes when I heard a "*chip*," and there were two "Golden-faced" Warblers—I call them that because all of a sudden I was suspicious that I might really be seeing Black-throated Green Warblers and might not be remembering what Golden cheeked Warblers looked like since I rarely saw them. I quickly snapped two pictures and raced back to the car to compare my pictures with those in the bird book, and I confirmed that I had seen Golden-cheeked Warblers.

---

BROWN-CAPPED ROSY-FINCH AND BLACK ROSY-FINCH

There are three species of rosy-finch, two of which are included in this book. All three are shaped somewhat like the common House Sparrow and are the same size (six and a quarter inches), with the primary coloration differences being on their heads. All of them are found in open areas, typically on the tundra near or on snow patches, where they hop along the ground. The Gray-crowned Rosy-Finch has a wide distribution, breeding from the Alaskan mainland and islands down to the northern US Rocky Mountains and wintering as far south as central New Mexico. The two species covered here have a much smaller range.

## Brown-capped Rosy-Finch *(Leucosticte australis)*

Global Population Estimate: 45,000, all in mountains of the western United States

Brown-capped Rosy-Finches are found in cold, mountainous areas. Even when the winter temperatures get down to −31 degrees Fahrenheit (−35

degrees Celsius), Brown-capped Rosy-Finches stay in the cold areas if food can be found. Generally they are only found in the high peaks in Colorado during breeding season and winter as far south as northern New Mexico. They do not have a formal migration pattern but just move to lower elevations when snow covers their breeding territory. They sometimes hybridize with Black Rosy-Finches, complicating their identification.

Male Brown-capped Rosy-Finches are a cinnamon-brown color with rose-tipped feathers on their wings, rump, belly, and tail. The female Brown-capped Rosy-Finches are a pale variation of the males' coloration. Brown-capped Rosy-Finches differ from the adults of the other two rosy-finches in that they do not have any plain gray areas on their heads.

The nests of Brown-capped Rosy-Finches are built in protected areas on cliffs, rock slides, or old buildings or in caves above the timberline, often near snowfields or glaciers. They eat seeds and insects in their rocky and snowy mountain territories and are often found on snowfields and grassy areas.

While there are no estimates of earlier population trends for the Brown-capped Rosy-Finch, recent data indicate a decline in the species, and therefore it is most likely that there were more Brown-capped Rosy-Finches before human activities, such as skiing, increased in the mountains.

Christmas Bird Count (CBC) results indicate that Brown-capped Rosy-Finches have been steadily declining in numbers, perhaps as much by 30 to 50 percent over the last thirty years in areas covered by the CBCs.

Rocky Mountain National Park supports a breeding population of perhaps at least one thousand Brown-capped Rosy-Finches. As with many of the species in this book, the remote areas where Brown-capped Rosy-Finches are found do not lend themselves to easy population studies, and therefore more information is needed on the birds' requirements, especially during the nonbreeding season. There is special concern that in particularly snowy years these finches may have difficulty finding food.

There is a need for additional data, such as through CBCs, to help determine the status of the Brown-capped Rosy-Finch. In addition, sightings of these finches should be reported on eBird. There are volunteer opportunities if you are in Sandia Crest in the winter, for example, to help fill feeders or leave donations of seeds, and to help at Crest House by telling customers about the rosy-finches. If you live in the mountains of Colorado or New Mexico (and do not live in a place such as the Wilderness Area, adjacent to Sandia Crest, where feeding is prohibited), you might consider providing bird seed, particularly in the winter and whenever snow covers the ground.

A Personal Note: My experience with Brown-capped Rosy-Finches is limited to seeing them in winter at Crest House at Sandia Crest in northern New Mexico. Rosy-finches of all three species winter there, at the very end of the road that ascends Sandia Crest (elevation 10,678 feet) (http://www.rosyfinch .com/). Crest House is a private business on land leased from the Forest Service. Even though feeding the birds is not allowed on the surrounding land, visitors can feed rosy-finches at Crest House in the winter. When I went to Crest House in 2008, I made the winter drive up the snowy mountain road, worried that my car was going to slide all the way back down or go over the edge. The trip was worth it, however, and all three rosy-finch species were there eating at the feeders.

## *Black Rosy-Finch (Leucosticte atrata)*
Global Population Estimate: 20,000, all in western US mountain areas

Black Rosy-Finches are basically black with pink on their wings and breast and a light gray band extending from the eye around the back of the head. The more common Gray-crowned Rosy-Finch is mostly black but also has gray on the back of the head, and this gray area covers most of the side and back of the head in two of the three populations of Gray-crowned Rosy-Finches.

The breeding range of Black Rosy-Finches is on barren stretches high in the mountains of the northern Great Basin (northeastern Nevada to southwestern Montana). Nests are built in a crevice or hole or on vertical cliffs. In winter the birds move down the mountains to more hospitable habitat, and in some cases they fly as far south as Colorado and New Mexico, where they often roost in abandoned buildings and other built structures. Black Rosy-Finches eat grass and weed seeds and insects, if available.

As with Brown-capped Rosy-Finches, there are no estimates of older population trends for Black Rosy-Finches, but there is evidence of a decline in the species. Although data are limited, Black Rosy-Finches appear to have been declining since the 1970s. The drop in numbers, together with the limited range of Black Rosy-Finches, is the reason for concern. As with Brown-capped Rosy-Finches, there is a need to study the population numbers of the Black Rosy-Finch, particularly as they relate to human activity on its breeding range, as well as to other possible threats.

As recommended by the Montana Partners in Flight Bird Conservation Plan, there needs to be a monitoring program at the high elevations where Black Rosy-Finches breed, and there also need to be actions to protect slightly lower elevations adjacent to the species' breeding areas.

As with the Brown-capped Rosy-Finch, there is a need for additional data. The National Audubon Society recommends that if you sight any of these finches, the sightings should be reported on eBird or at least to a local bird group. Volunteer activities, such as at Sandia Crest, and feeding the birds during winter are ways to help the Black Rosy-Finch.

A Personal Note: As noted above, I have seen Black Rosy-Finches at Sandia Crest in northern New Mexico. My first sighting, however, was in their breeding range in the Colorado Rockies. I was on a group field trip from Denver to Salt Lake in June. After scanning the snowfields with telescopes, we eventually located a distant Black Rosy-Finch sitting on the snow. My later winter sightings in New Mexico, often only a few feet away through an observation window, were much more satisfying.

## GOOD BIRDS VERSUS BAD BIRDS

Before I give some hints on how to help "everyday" birds, I need to mention that to most of us not all birds are equally "good." Most people who care about birds have a different level of caring about different bird species. Thus, if they put out a bluebird house for Eastern or Western or Mountain Blue-birds, they really want bluebirds to nest in it, not House Sparrows. If they put

out sunflower seeds for cardinals, they usually are less happy if it is all eaten by cowbirds and blackbirds and pigeons.

One of the reasons that certain bird species are deemed "good" birds is that these species are less common and more difficult to find than are other species. Of course there is a continuum—from birds that are very, very common, to those that are somewhat common, to those that are not easily found, to those that are rare, to those that are seriously "in trouble." There are many possible reasons that some species may be uncommon in a particular area, with one of the most likely reasons being that the habitat in that area is less suitable for that uncommon species than it is for the more common species. Thus, as forests are cleared for new homes, forest-dwelling birds such as Ovenbirds and many other warbler species lose their preferred wooded habitat. Similarly, when a wetland is drained, marsh birds such as rails and Marsh Wrens lose their homes. Birds that are more adaptable to changes in habitat, especially to environments containing lawns and homes and fields, such as House Sparrows, European Starlings, Great-tailed Grackles, and Brown-headed Cowbirds, generally do not suffer when native habitats are removed or substantially diminished, and in fact they often flourish. Improving the habitat by returning it to its natural state, or to a state as close as possible to its natural state, gives the "good" species that are native there a better chance at competing and surviving.

For most of us, it is possible to strike a balance so that less common birds, particularly the threatened birds, have a better chance of surviving and reproducing while the more numerous birds, which at times might seem to be taking over all the available habitat, don't receive much special attention. While at times we might think that we should just live and let live—let nature take its course without our affecting the end result—our being here at all has already affected the result. Our homes and our cities and our fields are standing where birds and their habitat used to be. Some birds love our cities and flourish there (e.g., grackles), while others require the wide-open spaces or the woodlands or the swamps that are disappearing. Helping the birds that our civilization has adversely affected and trying to control those that threaten to overrun our cities seem in my opinion to be good ways to reach a balance. This section of this book provides ideas on how to help strike that balance.

## WHAT WE CAN DO FOR THE MORE COMMON BIRDS

When we learn about rare birds and the huge difficulties that some of them are facing, it may seem like there's really nothing that most of us can do about their problems that will make a difference. Of course, for some species that may be the case. Each of us, however, can still do something of value to help birds. While no one can be all things to all birds, everyone can try to make a difference. Even some of the more common birds that fly through or over our neighborhoods or nest near where we live often have difficulty finding what they need to survive, to nest, and to prosper, especially if the habitat has become less favorable for birds of the species. Most of us CAN do something to help these birds. We might be surprised that, when we improve our local habitat for the less common birds, we might actually see birds there that are rare or in trouble.

Birds, like humans, have basic requirements, which vary only somewhat by species. These basic needs of the birds are food, water for drinking and bathing, shelter, and a safe nesting spot. Depending on the time of year and the bird species, what is done in your yard or a local nature area near you may make it possible for birds to find what they need in the midst of the human-changed environment.

The following pages, which refer primarily to what you might do in your yard for the birds, can be extended to any other area in your neighborhood, city, county, state, or beyond where you have the possibility of improving the habitat for birds. This potential area of activity includes nature preserves and other natural areas where the owner or manager of the property is willing to allow the area to be maintained or improved for the benefit of birds.

These pages are meant to encourage you to figure out what you and others might do to help the "good" birds meet their needs in spite of what humans have done to reduce or destroy the environments that the birds require to flourish. In addition, you can help birds survive when nature itself causes habitat reduction or destruction, such as in a prolonged drought or severe snowstorm or other environmental catastrophe or when the "good" birds are facing stresses, such as preparing for long migratory journeys. Much more detailed information on what you can do to help the birds can be found in a large number of books and periodicals and online, but this section can get you started.

Whether or not you feed birds year round in your yard, try if at all possible to put out bird seed during winter in cold climates, particularly when the weather is bad, such as when snow covers the normal feed sources and birds often become stranded. Similarly, when there is a drought in your area, putting out water for the birds can mean the difference between life and death for them. Putting out hummingbird feeders during their migration in late summer and fall can provide the hummingbirds with a place to stop and put on a few more milligrams of fat before they attempt their hazardous crossing of the Gulf of Mexico or arid land areas.

Any person who has outside yard or porch space and who lives where there could be birds (which is almost everywhere) can do something for the birds in their area to help meet the birds' requirements. Where you live and the amount of space that you have will be important in what you decide to do for the birds around your home. Below are some ideas to get you started. Generally I have not provided brand names or specific product recommendations, but rather I encourage you to go to stores and find websites that sell bird products to see what is available that might be suitable.

It is important to stress that whatever you do in your yard or local nature area to help the "good" local birds may be very important in helping a rarer "in-trouble" bird that comes to your area.

### Water

Unless you live very near a lake or other natural freshwater feature that has shallow areas from which birds can easily drink, it is likely that the most important thing you can do for birds is to provide water, especially in dry seasons or times of drought and especially in arid locales or where water has been removed or reduced by human activities.

There are many types of products that will serve birds in need of a drink or a bath. The goal should be to emulate as much as possible the ideal natural birdbath. Thus, it is best if the maximum water level in the birdbath is relatively shallow. If birdbaths are deeper than a couple of inches, small birds risk drowning. Deeper birdbaths are especially dangerous if they do not have a shallow area around the edge or if they have a steep edge that rises significantly above the level of the water. Placing small stones or other perches in the water provides a way for smaller birds to have better, safer access to the water.

It is important to clean the birdbath and refill it with fresh water on a regular basis. While some natural algal growth in the birdbath may be harmless,

diseases can be spread if birdbaths are not cleaned regularly. Some birds, such as grackles, regularly dunk their food into the water of a birdbath. This natural behavior can dirty the water with food particles, which can then rot and provide food for microorganisms to grow in the water. Also, the more birds that are using the birdbath for bathing, the more likely it is that bird feces will pollute the water.

Although a simple pan would make an adequate birdbath, adding water features to a birdbath would make it a major attraction for birds because, as in the natural environment, a trickling brook often attracts birds. Having vegetation in or near the water feature makes the birdbath more similar to a natural habitat for the birds.

In our North Texas yard, which was not near any lakes or streams, we provided a birdbath and also dug a hole for a prefabricated pond insert that we waterscaped with various water plants. We also dug and lined a second hole with plastic, perforated the plastic in a few places, then refilled the area with dirt to make a bog, adding a variety of wetland plants. This manufactured marsh attracted birds, but even better at attracting them was the addition of water dripping into the nearby birdbath so that runoff from it went into the bog. The dripping water and nearby bed crowded with bog-loving plants proved irresistible each year to migrating warblers such as Mourning Warblers and American Redstarts.

Because some birds are more comfortable on the ground while others are more comfortable perching above the ground, having a birdbath on the ground as well as a traditional birdbath on a pedestal can increase the number of bird species that come to your yard for water by providing a more varied habitat. It is also best if most or all of the birdbaths can be near shelter, such as a bush, so that bathing or drinking birds can more easily dart away into safety if they feel threatened.

If you live in an area where the temperature during some part of the year is below freezing, you can still provide water for the birds. Most birds prefer to drink liquid water if it is available and apparently only ingest snow when necessary. Of course liquid water is required if birds are to bathe. The simplest way to provide liquid water in winter is to have a birdbath that you can turn upside-down so that hot water can be poured over the bottom of the bath to loosen and remove ice; the birdbath can then be refilled with water on a regular basis. If the birdbath is near enough to an electrical outlet, there are also various electrical birdbath warmers that can be used, as well as solar devices that, at least in climates with sufficient winter sunlight and warmth, can keep the water from freezing.

### Food

Unless you live in a particularly arid area with little or no vegetation, it is likely that during much of the year there will be no need for you to supplement the diet of the local birds. When the weather turns cold, however, particularly if seed-bearing plants are covered with snow, birds will usually flock to seeds thrown on the ground or snow or placed in or on feeders. Even in the warmer months birds can benefit from the addition of food to your yard, especially if they are feeding young.

Most birds, particularly in winter, can be attracted to your yard and be satisfactorily fed if you put out their preferred foods—those that are the same as or similar to their natural foods or those that are particularly high in components needed by the birds. Stores that cater to birders often have information on what types of seed or other food will attract the different species of birds that come to the area. Of course, there are also many resources on the Internet and in books to assist you in determining what food to buy to attract birds. You can also experiment to see which birds come and partake of your offering. It should go without saying that whatever feeds you put out for the birds should not be spoiled or contaminated with deadly pesticides and that feeders should be kept clean, particularly if there are known or suspected bird diseases around (e.g., House Finch disease, also known as avian conjunctivitis).

One useful website is provided by the Cornell Lab of Ornithology at www .feederwatch.org. It provides information on what common feeder birds like to eat, where they like to eat it, and what type of feeder works best for them. The site allows you click on a bird picture and see its food and feeder preference, or you can select your region and a food and/or a feeder type and then see what birds you might attract.

Below is general information on some possible food types to consider trying in your yard.

Fruit. There are many birds that eat fruit. American Robins, Cedar Waxwings, Northern Mockingbirds, and bluebirds, for example, are attracted to berry-bearing bushes and trees (e.g., mulberry), apples, and other fruits, especially if the plants hold their fruit into the fall and winter.

Orioles, House Finches, and sometimes woodpeckers can be attracted to orange halves skewered on branches or feeders, as well as to grape jelly placed in feeder depressions.

Mealworms. Mealworms for use in feeding birds are sold in dried form or living in a medium such as sawdust, and mealworm feeders can be purchased. Typically such feeders have a shelter over them to keep the worms dry. To keep them from drowning if it rains the feeders have holes in the bottom so that water can drain out. Birds that eat insects in the wild, such as bluebirds, wrens, woodpeckers, and robins, can be attracted to mealworms, particularly if the birds are nesting nearby. After I moved to Alaska, I found that Steller's Jays, while foraging for food after the snow fell, eagerly gorged on dried mealworms that I had placed in the base of an old hummingbird feeder.

Seeds. One of the most confusing things about feeding birds is the huge variety of seeds that are sold for that purpose. Sunflower seeds are almost universally a good choice when starting to feed birds. Many wintering birds, such as Northern Cardinals and grosbeaks, are attracted to sunflower seeds, both black-oil sunflower seeds and striped sunflower seeds. Striped

sunflower seeds have harder shells and are best suited to birds that have sturdy seed-cracking bills. Feeding striped sunflower seeds therefore can be a way to discourage blackbirds and House Sparrows. Cardinals also eat white safflower seeds, as do doves, chickadees, and other small birds.

To favor the generally rarer woodland birds, especially in agricultural and suburban areas where overly adaptable House Sparrows, blackbirds, and cowbirds are crowding out the less common "good" birds, one should avoid putting out seed mixtures that contain varieties especially preferred by the less favored species. White proso millet seed is one example of a type preferred by less favored birds.

Similarly, while cracked corn spread on the ground is likely to attract pheasants, turkeys, crows, and many other birds, these are birds that often flourish in cleared land habitats and do not usually require feeding. The feed preferred by these birds also attracts cowbirds and is less likely to be eaten by birds that are threatened or in need of protection. It is also important to keep from

oversupplying corn; uneaten corn may spoil and provide a favorable environment for growth of mold, such as the type that produces poisonous aflatoxins.

Shelled and unshelled peanuts are also attractive to birds. As with corn, it is important that peanuts not become wet and moldy, as aflatoxin production is likely to occur in wet peanuts. As may be the case when too many pest birds are present, an overabundance of squirrels in your yard may require changes in where the peanuts and other feeds are placed.

Thistle seed (Nyjer). Most of the bird species that eat thistle seed are birds that can experience food shortages during the winter in areas where their usual seed-food is covered by snow or in areas with limited seed availability where these species winter in huge numbers. Seed-eating birds such as American Goldfinches and other finches, as well as Pine Siskins, often prefer thistle seed in the wild. While often called "thistle seed," sterilized Nyjer seed that is similar in appearance to thistle seed but even more attractive to finches is often sold by bird seed companies. Nyjer seed, a black bird seed grown in Asia and Africa, is very high in oil and calories. Sterilized Nyjer seed has the added advantage that it does not produce thistle plants when the seeds fall to the ground in your yard. Because of the very small size of Nyjer seeds, special feeders are used to dispense them. Examples include tube feeders with perches and tiny holes above (or sometimes below) each perch and mesh feeders designed so that finches cling to the mesh and pull the tiny seeds through the holes in it.

In order to optimize the attractiveness of your yard and provide feed for the maximum number of species, it is best to provide a variety of seed types, as well as to provide the seed in different types of locations, such as on the ground, on platform feeders, and in different styles of feeders. While some birds, such as chickadees, will come to get their preferred black-oil sunflower seeds wherever you put them, other birds can be more particular about where they will come to eat. The more your feeder contents and location are designed to correct habitat deficiencies and to imitate a natural food source that is preferred by a "good" bird species in your area, the more likely that birds of that species will find and feed from your feeder.

Bird seed feeders. There are many types of feeders for bird seed, but most feeders fit into one of the basic types: feeders for placement on the ground (typically low platform feeders that allow ground-feeding birds to find and eat the food); platform feeders placed on pedestals (for birds accustomed to feeding aboveground, such as in bushes); tube feeders; and large and small hopper feeders (for birds that perch on tree branches). In addition, many bird seed companies offer seed balls, seed wreaths, and other forms of seed compacted together, and these foods should be placed so that birds can feed on them in safety. As with types of seeds, it is often useful to experiment with different types and locations of bird feeders to see what the birds in your area prefer, especially the "good" birds that may already be in your area but be in need of food in the habitat around you or the "in-trouble" birds that may be migrating through the area or attempting to breed in the area and are having difficulty finding food.

Suet and related products. The "suet" used as a bird food originally was the fat around the kidneys and loins of cows and sheep, but this word has come to mean simply any fat trimmings from beef or other meats. Animal fat is very digestible by birds and in winter can meet their need for protein that would normally be met by insects, so it can be a valuable addition to their diets, especially in winter. In colder climates, suet placed in wire-cage suet feeders will last many months. In warmer climates, other high-fat compositions on the market can be used instead of animal fat, which becomes rancid at higher temperatures and can melt and drip out of the feeders. Many people have developed home recipes for suet mixtures, often containing peanut butter and other fats, as well as seeds and fruit, and birds find such mixes attractive.

When I lived in North Carolina, I kept my suet feeders filled with a peanut butter mixture, and they attracted not only the usual suet-loving birds (woodpeckers, chickadees, nuthatches) but also wintering Baltimore Orioles and American Robins. Unfortunately, starlings also loved that suet mix, so in order to discourage the starlings I periodically used the less "tasty" commercial suet blocks because they lasted longer in my yard and gave the "good" birds a chance at finding the suet.

Hummingbird food. Hummingbirds in the wild eat tiny flying insects and nectar from nectar-rich flowers. Planting for hummingbirds is discussed elsewhere in this section of the book. If you are going to provide the maximum benefit for migrating or breeding hummingbirds, you need to consider putting out hummingbird feeders containing sugar water in addition to doing planting for hummingbirds (of course, as I learned when I moved to Alaska, in some areas feeders, including hummingbird feeders, should not be used in summer, since they attract bears). Hummingbird feeders generally have red color on them somewhere to attract hummingbirds, which (away from feeders) usually feed from red flowers that are high in nectar. The composition of sugar water in hummingbird feeders approximates flower nectar (typically one part sugar

dissolved in four parts water) and, if available, is eagerly consumed by most hummingbirds. Most of the birds that are common and not in need of our help are not likely to drink sugar water from hummingbird feeders.

It is particularly important that hummingbird feeders be cleaned often and refilled with fresh sugar water. Sugar water should not be allowed to become cloudy due to microorganism growth or to smell like vinegar or yeast. If any of these conditions occurs, the spoiled sugar water needs to be dumped, the feeder cleaned, and fresh sugar water added. In hotter climates it may be necessary to do this every two to three days.

If you have hummingbird feeders in your yard during migration but no hummingbirds have found your feeders, it can be helpful to add more red color to your yard. I have in the past taken a red plastic dinner plate and anchored it in the yard near a feeder to catch the passing hummingbirds' attention. My theory is that it can't hurt and might improve the hummingbird situation in the yard.

Planting

There are many reasons to plant annuals, perennials, shrubs, vines, and trees to help the birds by simulating as closely as possible the natural habitat, particularly if your yard has few or no plants taller than lawn grass. While any plant that provides possible cover and shelter for birds or provides food for birds can be planted in your yard, it is strongly recommended that only plants that are native to the area be cultivated. Not only are native plants more likely to grow well in your yard and be the kind of plants with which your birds are familiar, but they are also less likely to be invasive and take over your yard, and your neighbors' yards.

Evergreen plants, both conifers and nonconifers, can provide nesting sites and shelter for birds all year, and they are thus valuable additions to a yard. Also, the seeds found in the cones of many conifers are attractive food sources for northern birds such as crossbills and Pine Siskins.

Plants that provide multiple benefits, such as seeds, flowers for nectar, or fruit, in addition to shelter are great boons to the birds that come to your yard and will enhance your own enjoyment. Sunflowers, for example, are easy to grow and provide some of the seeds that birds love best. Trees and shrubs that bear small fruits such as cherries and berries can prove to be veritable bird magnets when the fruit becomes ripe. Of course, the birds clearly will compete with any human interest in picking the fruit, and the food interests of birds and humans will not always be compatible.

There are numerous plants that produce flowers that are both attractive to hummingbirds and beautiful additions to a yard. Most seed and plant catalogs, as well as many plant retailers, provide information on which plants produce nectar and attract hummingbirds. Usually, but not always, such plants have red or purple flowers. Many plants that attract hummingbirds have the added benefit of also attracting butterflies.

Flowers in the *Salvia* genus, which includes both annuals and perennials, are particularly attractive to hummingbirds, but there are many other good plant choices for hummingbirds. Other often-recommended plants for attracting hummingbirds are those in the genus *Monarda* (for example, bee balm) and *Anisacanthus* (favored by a hummingbird that visited my North Texas yard, though the plant is not native to the region), as well as fuchsia, trumpet vine, hollyhocks, columbine, and penstemon. There are many websites that provide information about choosing plants for hummingbirds; the Hummingbird Society has a good example (http://www.hummingbirdsociety.org/hummingbird-flowers/). While many of these plants are not native to North America, the hummingbirds that they attract are mostly likely used to feeding on these same plants in Central and South America and the Caribbean during our winter. In my experience, the few non-native hummingbird plants that I have planted have not been problematic by becoming weedy and out of control.

If you have a wet area in your yard, or want to create one by making an artificial bog, plants that grow well in wet soil are good choices. In the little bog that I made in my North Texas yard, I planted not only irises but also annuals that produce hummingbird-attracting flowers, such as cardinal flower (*Lobelia cardinalis*).

Depending on where you live and how much space you have available, you can consider going beyond the planting of a few trees or bushes and developing an entire backyard habitat tailored to and suitable for the "good" birds that live in or migrate through your area. Thus, in north-central Texas, where hummingbirds (mostly Ruby-throated and Black-chinned Hummingbirds) sometimes nest but through which they always migrate in spring and fall, I decided to plant as many hummingbird plants as I could. I planted both annuals and perennials and observed to which plants the hummingbirds went and which plants survived the hot summers and sometimes amazingly cold winters. Having a great variety of plants in diverse locations in the yard (as well as numerous sugar-water feeders) allowed multiple hummingbirds to feed relatively peaceably in spite of their feisty attitudes toward each other. The nectar from these plants also augmented their diets as they worked to put on fat for their long migratory voyages.

After I moved to South Dakota, where our home was adjacent to a large field, my backyard habitat became more of a prairie-edge habitat, with my

contribution to the birds being mainly water and seeds, along with seasonal hummingbird-attracting flowers and sugar water. I visited a birder in central Minnesota who has a large enough yard and sufficient energy and devotion to create an actual small prairie with native prairie grasses and flowers. It is a delight to see, and I am sure it is a welcome spot for many prairie birds.

### Brush Piles and Brush Areas

I learned about the value of brush piles after we cut off many low-hanging branches of our backyard pecan tree in North Texas and I did not get all of the branches out to the curb for collection. The next winter we had sparrows come and winter in our yard, remaining much of the time in the downed pecan branches. Some of these birds I had never seen in our yard before. They were birds that preferred a brushy habitat that is not often found in a typical suburban setting, but my brush pile had given them a hiding place in the midst of a subdivision. Most people who try to make their yard beautiful shudder at the thought of a messy brush pile on the premises, but if there is some corner of the yard where the presence of brush is acceptable to you, I strongly recommend it as a place for native sparrows to linger.

In South Dakota, where I had few trees or branches from which to make brush piles, I have found that rescuing "used" Christmas trees in January and bringing them to my yard made a brush pile that provided a place of retreat for Dark-eyed Juncos and American Tree Sparrows.

### Fallow Areas

Another way to attract birds to a yard, especially in winter, is to refrain from removing plants in fall after the growing season. The yard will then more closely resemble an uncleared native habitat. Areas with, for example, old seed heads of coneflowers and other composites, as well as areas where grass is allowed to grow and set seed, provide a place for birds to feed in addition to your feeders and also serve as shelter for ground-loving birds. Depending on the type of plant, it is not always necessary to cut plants back after the growing season if you do not mind seeing brown plants in your yard. Any remaining vegetation can provide shelter for birds, even if it does not provide seeds.

### Artificial Shelters and Nest Sites

In winter and inclement weather, many birds seek shelter. While increasing the vegetation in your yard is generally the best way to provide shelter, some birds also will go into artificial shelters. Usually these are the same birds that during nesting season will use a nest box instead of a hollow in a tree for nesting. Thus, chickadees, which often roost in groups, will go into either holes in trees or comparable shelters. Screech-owls will also roost in nest boxes. I did learn in North Texas, however, that putting up a screech-owl box does not guarantee a screech-owl but can provide a happy home for squirrels.

Most birds that nest in cavities will also use nest boxes, if present. Preferably, such nest boxes should be properly designed for the desired birds so that the birds will be attracted to a place that is similar to a natural site for them. A proper nest box design will also help cut down on invasion by House Sparrows and starlings, which also nest in cavities. Details of proper design for various species include the hole size, box size, and location of the nest box. Such information is available on the Internet, as well as in books and from members of local bird clubs.

Bluebirds, which often can be induced to live in or near neighborhoods, are undoubtedly the most commonly sought nesting birds. The North American Bluebird Society (http://www.nabluebirdsociety.org/) provides information on the different species of bluebirds and how to attract them by, among other things, providing nest boxes for them. There are also many communities and local groups that have established and maintain bluebird trails with bluebird nest boxes. Natural holes in dead trees that bluebirds use are those without any branchlets nearby, and thus bluebird nest boxes should not have perches next to the holes. Perches, which bluebirds and swallows do not need to enter the nest box, enable predatory birds, as well as House Sparrows, to land and easily enter the box. Larger predators can use them to reach through the opening in the box and remove eggs and baby birds.

If you have trees in your yard that already have cavities, you may be lucky enough to have a woodpecker nest in a cavity. If a tree dies in your yard and the yard is big enough (and informal enough) that you can allow it to remain (possibly topped off to avoid falling dead branches), you can watch a woodpecker habitat develop as woodpeckers drill into the trunk.

Some birds routinely take advantage of anything that will support their nesting materials. I hear stories all the time about House Finches nesting and raising babies on wreaths hung on doors and in hanging flower pots, as well as about House Wrens and other wrens nesting in anything with an opening (such as boots, bottles, a knothole in a garage wall, and the like). American Robins will place their mud-lined nests on platforms under the

eaves of buildings. Of course, some birds nest in places that are not right for them, such as the Blue Jays at my parents' house in Wisconsin that tried to place their stick nest on a platform under the eaves that was meant for robins. The jays could not understand that when the sticks on the platform got to a certain height, any additional sticks brought to the nest fell to the ground beneath the platform. By the time the jays gave up on building the nest, the pile of sticks below the nest platform was about three feet high.

It is also often helpful to birds that are beginning to make their nests if you put out nesting materials, particularly if there are no other obvious sources of nesting materials. Such nesting materials can be placed in a wire cage, such as a feeder designed to hold suet blocks, and hung in your yard. For example, Tree Swallows will not only eagerly snap up chicken feathers to put into their houses but also are likely to entertain you as they fly after feathers that are blowing in the wind. Fine grass clippings and small twigs can also be put out. In Texas, when I dug up sod to plant a new garden area, Blue Jays would come down as soon as I left the yard and would pull up grass and roots from the excavated sod and use the material in their nests.

While most people think of little birds like bluebirds, wrens, and swallows when they consider putting up birdhouses, if you live near a lake or are working to improve a habitat surrounding a lake, you might consider putting up a large house for Wood Ducks. At Canyon Lake in Rapid City, South Dakota, Wood Ducks used to be rare at best, but when Wood Duck houses were put up near the lake edges, the Wood Ducks came, nested, and raised babies and are now relatively common at the lake in the summer.

If you live in an area where Purple Martins are found, your yard may be the perfect place to put a Purple Martin house. Martins nest in colonies, so martin houses usually have six or more "apartments" in which they can nest. The closer a martin house is to water, the more likely the martins, which often hunt insects over water, are to find and use the house. Also, the martin house should not be placed in a heavily wooded area; instead, situate the house so that at least one side opens onto an unobstructed flyway the birds can use when going to and from the house. One very nice thing about Purple Martins is that they do not appear to become particularly nervous when people are nearby, and they can thus be easily observed flying in and out of their nests to feed the nestlings.

### Chimney Swift Roosts

If a habitat has trees and bushes, there is usually a place for most small birds to roost at night. Chimney Swifts, however, require a very specialized spot

to roost. Before humans began regularly cutting down hollow trees, Chimney Swifts roosted inside them, as well as in caves. Back when cities featured many large chimneys, such as in schools and other public buildings, swifts used them for roosting spots. As older buildings were replaced by buildings without tall chimneys or tall chimneys were covered with wire to prevent the birds from entering, the Chimney Swifts no longer had available roosting habitats. Chimney Swift towers were designed to meet the need for roosting spots. Designs for such towers can be found on the Internet (for example, at http://mn.audubon.org/chimney-swift-towers). Typically, the towers are eight to twelve feet tall and have an interior surface to which the swifts can cling. They also have various design features to prevent predator problems. Local bird groups in areas where there is a need for such roosts have worked with their communities to put up Chimney Swift towers.

### Things NOT to Leave Unattended in Your Yard (or Elsewhere)

Chemicals. It is important to minimize and, if possible, eliminate the use of chemicals in your yard because pesticides can kill birds that eat the dead insects or the seeds from weeds treated with herbicides. It is also important to remember that pesticides meant for "bad" animals (e.g., mice, rats, pigeons, grackles, etc.) will also poison a "good" bird and can kill a cat or dog that eats the dead or dying animal. In addition, chemical fertilizers and other chemicals can run off into streams and lakes and cause algal blooms or kill animals or other organisms in the waterways. Even where there is no nearby waterway, chemicals placed on the soil can contaminate plants and plant seeds that are there and can also leach into the soil and then into the groundwater. If some type of chemical is to be used anywhere outdoors, "bird-friendly" products should be used whenever possible. After moving north (to South Dakota and then Alaska), I found that there was such a thing as a bird-friendly ice-melt product that I could use on the icy patches on our driveway and sidewalk rather than the chloride compounds in the unfriendly products. Basically, any product that is not naturally in your yard already and known to be a good, harmless thing is suspect and needs to be evaluated before adding it to your yard.

Cats. This topic is very controversial for many people, especially cat owners, of course. It is well known, however, that even well-fed, happy, healthy cats have a strong instinct to catch any living thing they can, including birds. Unrestrained cats will notice places where birds tend to gather, feed, roost, or drink and be attracted to the bird gatherings. Therefore, in order to reduce the carnage and to protect the birds, cats should not be allowed to roam uncontrolled outside in your yard or into the wider neighborhood. Most cats are quite happy watching the birds at the bird feeders from inside the house. Other cat owners who wish to allow their cats outdoors have found that an enclosed wire run for the cats can be satisfactory.

Liquids. Other than natural freshwater streams or reservoirs that are regularly flushed and refilled, open containers of any liquid generally should not be in your yard. Such containers can be hazards to birds, which can drown in them. If the water has oil floating on it, birds' feathers can be polluted, just as in a toxic oil spill or at an industrial waste site. Toxic liquids such as antifreeze can of course kill any bird (or other animal, such as your dog) that drinks them. Even open containers of any size that hold water for an extended period, for example, after a rainfall, can be the breeding ground for mosquitoes if left unattended. If you have a pond in your yard, there are "natural" products you can use to at least control mosquito populations without harming birds or other animals. Generally, these products contain *Bacillus thuringiensis*, which is a bacterial species that kills the mosquito larvae in the water but does not hurt other animals, including those that might eat the larvae.

Plastics and other litter. While we usually think that the reason not to litter is to keep the environment looking nice, it is important to be aware that some harmless-looking things can be dangerous to birds, whether such things are thrown out along the highway, left on the beaches, or put into garbage bags to be taken to a landfill. Birds of any size at a landfill can become trapped in the holes in a discarded plastic six-pack holder for soda, and birds at the beach can become entangled in nylon fishing line. Garbage and litter that is nondegradable and in which birds can conceivably become trapped in any way needs to be either cut into pieces that are not dangerous or enclosed as permanently as possible in a sturdy container.

### Hazardous Structures

While structures are not typically a problem in an average-sized yard, in larger areas through which birds migrate, including ranches, fields, park areas, and the like, it is important that large structures not be erected in the flight pathway. Large moving structures, such as windmills, are particularly hazardous along migratory pathways. There have been many studies by environmental groups, as well as by landowners, of the hazards presented by such structures, with the conclusion always being that such structures can be deadly to many birds. In the West, such structures are especially hazardous to eagles, which are drawn to their death in the eddies of air around windmill blades.

Tall buildings with glass windows that reflect the sky, as well as tall structures with lights, regularly cause many songbird deaths in cities. Anyone with a large picture window, or with smaller windows near a bird-feeding area, can attest to the fact that birds regularly fly straight at a glass pane, not being aware that it is a hard surface. There are continued efforts to design structures to minimize bird collisions, for example, by reducing the amount of glass surface, by angling the windows to reduce reflection, or by placing images or lines in or on glass surfaces. For backyard bird feeders, removing feeders from the immediate proximity to glass windows can make a difference in the number of birds that crash into the glass, as can putting numerous opaque surfaces on the window.

### ABA CODE AND THE LAW

Most birdwatchers are familiar with the American Birding Association (ABA) and its Principles of Birding Ethics. Whether or not you go out to watch birds and whether or not you decide to try to improve your yard or other habitat for the birds, these principles are very important for you to know and follow (http://www.aba.org/bigday/ethics.pdf). In summary, these are principles that all of us should follow to promote the welfare of birds and their environment. By doing so, people can support the protection of important bird habitat and minimize disturbance to birds, respect the law and the rights of others, and ensure that feeders, nest structures, and other artificial bird environments are safe.

Under the various laws discussed in more detail earlier this book, it is illegal to collect or even pick up dead birds, feathers, nests, or bird eggs without the appropriate permits. Such collecting is unethical and likely to be a problem for all birds, and it is also against the law to kill or harass birds. When birds are nesting, it is critical to stay away from their nests. Don't pull away leaves and branches so you can see or photograph the nest because doing so exposes the nest to the elements and to predators.

### Baby Birds and Injured Birds

If you find a baby bird that cannot fly and is in an unprotected area away from the nest, you should try to return it to its nest or to an area away from predators that are near the nest, for example, to a nearby bush. It is generally believed that birds cannot "smell" any trace of humans if their young have been handled. If you find a bird that is injured, you should try to find someone licensed to care for wild birds. In many cities, there are licensed rehabilitation people ("rehabbers") or veterinarians who can help.

### Photographers

Many birders do not like bird photographers because they are perceived (often rightly) as not being as concerned as birders about the welfare of the birds being photographed or about the desire of birders to see the bird without scaring it away. Some birders, including myself, are also photographers and ARE concerned about the birds and about other birders seeing the bird too. All bird photographers need to be aware that their "need" for a better picture can result in harassment of the target bird. They need to be aware of others who are trying to see the birds, so photographers should not approach birds closer than others are doing or do anything that is stressful or harmful to the birds.

### VOLUNTEERING

It is very likely that you live in or near a community that has a park, a greenway, or some other public land that could be preserved, protected, or enhanced for the benefit of the birds. See if you can help! Whether or not you have a yard to improve for birds, you might also consider volunteering for local organizations, parks, and wildlife areas to help these other areas be improved as bird habitats and stopping-over areas, especially for the "good" birds. If you do have a yard, you might improve it as well, to provide possible benefit for an "in-trouble" species. Volunteer opportunities include helping in cleanups, providing information to the public, or keeping feeders stocked, invasive plants eliminated, and paths cleared. You can also help by keeping bird records (regular surveys) in local wildlife areas so that those who manage and own the areas can be made aware of good and bad trends in bird numbers and thus know how to proceed to make the area better for the birds.

The more informed you become on birds' habits, habitats, populations, and problems, the more you will be able to help others become informed. Speak up. Support wild habitats—parks, backyards, and wild areas—and whatever can be done to manage and improve them.

### A PENULTIMATE NOTE ABOUT HELPING BIRDS

While this section of the book stresses providing food and water for birds, it is true that some "bird people" believe that feeding birds, especially during migration, disrupts the birds' normal migration patterns and keeps them from completing their migration on time. Most birds, however, stop along their migratory route to take in the calories that will enable them to complete their migration. Your bird feeder is just making the food more available for them. The migratory urge is so strong in most birds that the availability of food is unlikely to make them forget about migrating. For birds that are hungry or in poor health, providing food and water may make it possible for them to continue on their journey.

For the infrequent case in which a bird arrives during migration and stays rather than continuing on the migration, if you keep providing food, water, and shelter until the bird moves on, I do not see that you have done it any harm. For example, during ten winters in North Texas a female Rufous Hummingbird (a species that often wanders far afield from its normal migration route), which was possibly but not necessarily the same bird all those years, came to my yard, drank my sugar water, and stayed the winter. Most years she stayed from late summer (August) or early fall until early April, year after year, providing enjoyment for me and many others and drinking a large quantity of sugar water.

### A FINAL RECOMMENDATION

Throughout the text above, I have included many suggested courses of action. If each of us does some of these things, we can start to make a difference. But reading (and writing) is only a first step, a dipping of our toes in the water. There is so much more to know about the birds and so many more things we might do. I hope that you do not stop when you finish this book. Read it again, mark it up, then go out and do something. And keep reading. I particularly recommend Jeffrey Wells's *Birder's Conservation Handbook* (see Suggested Reading), which discusses "in-trouble" birds in more detail, with footnotes. Also, Laura Erickson's *101 Ways to Help Birds* provides many other suggestions for helping birds.

| Species | Global population, 2007 | Other population estimates |
|---|---|---|
| Trumpeter Swan | 34,803 | >30,000[1] |
| Gunnison Sage-Grouse | 2,000–5,000 | <4000[1] |
| Greater Sage-Grouse | 150,000 | 142,000 in 1998 (including Gunnison Sage-Grouse)[1] |
| Greater Prairie-Chicken | 690,000 | |
| Lesser Prairie-Chicken | 32,000 | 17,616[2] |
| Yellow-billed Loon | 23,500 | 10,000[1]; 17,000 (3,000–4,000, AK)[3] |
| Clark's Grebe | 15,000 | |
| Black-capped Petrel | 5,000 | <2000[1] |
| Bermuda Petrel | 180 | 142[4] |
| Ashy Storm-Petrel | 5,000–15,000 | 5,200–10,000[1] |
| Reddish Egret | 67,500 | USA, 2,000 pairs: 1,500, TX; 350–400, FL; 50–150, LA; a few, AL[5] |
| Ferruginous Hawk | — | 5,842–11,330, North America[5] |
| Yellow Rail | 17,500 | |
| Black Rail | Unknown | |
| Whooping Crane | 593 | 599[6] |
| Snowy Plover | 370,000 | 21,000, USA[1] |
| Piping Plover | 6,410 | 6,000[1] |
| Mountain Plover | 8,500 | 8,000–10,000[1] |
| Wandering Tattler | 15,000 | 10,000–25,000[1] |
| Bristle-thighed Curlew | 10,000 | 3,200 breeding pairs[1] |
| Long-billed Curlew | 20,000 | 20,000[1] |
| Hudsonian Godwit | 50,000 | No more than 50,000[1] |

| | | |
|---|---|---|
| Red Knot | 1,100,000 | 400,000, North America |
| Buff-breasted Sandpiper | 15,000 | 15,000[1] |
| Kittlitz's Murrelet | 24,000 | 9,000–25,000[1] |
| Xantus's Murrelet (Scripps's and Guadalupe Murrelets) | 5,600 | 5,600[1] |
| Craveri's Murrelet | 15,000–20,000 | 5,000 breeding pairs; up to 15,000–20,000 total[1] |
| Ivory Gull | 15,500–23,900 | |
| Flammulated Owl | 37,000 | |
| Spotted Owl | 15,000 | |
| Red-cockaded Woodpecker | 20,000 | 5,000[1] |
| Red-crowned Parrot | 5,000 | |
| Black-capped Vireo | 8,000 | |
| Florida Scrub-Jay | 6,500 | 4,000 breeding pairs[1] |
| Island Scrub-Jay | 9,000 | 12,500 individuals (7,000 breeders)[1]; 1,705–2,267 birds[7] |
| California Gnatcatcher | 77,000 | |
| Bicknell's Thrush | 40,000 | 50,000[1] |
| McKay's Bunting | 6,000 | 2,800[1]; 31,200[5] |
| Colima Warbler | 25,000 | |
| Kirtland's Warbler | 4,500 | |
| Cerulean Warbler | 560,000 | |
| Golden-cheeked Warbler | 21,000 | 9,600–32,000 (1990)[1] |
| Brown-capped Rosy-Finch | 45,000 | |
| Black Rosy-Finch | 20,000 | |

Note: All global population estimates in the middle column come from Audubon Watchlist 2007, http://birds.audubon.org/2007-audubon-watchlist.

[1] American Bird Conservancy Watchlist Species Accounts, http://www.abcbirds.

[2] "Range-Wide Population Size of the Lesser Prairie-Chicken: 2012 and 2013," report prepared for the Western Association of Fish and Wildlife Agencies, 2, http://www.wafwa.org/documents/Final_08.13.2013_LPC_Popestimation.pdf.

[3] Natural Resources Defense Council, http://www.nrdc.org/wildlife/habitat/esa/alaska04.asp.

[4] Birdlife International, http://www.birdlife.org/datazone/speciesfactsheet.php?id=3910&m=1.

[5] Birds of North American Online, http://bna.birds.cornell.edu/bna/.

[6] International Crane Foundation, 2014 estimate of total population (wild and populations being established), https://www.savingcranes.org/.

[7] Smithsonian Migratory Bird Center report, 2012, Ecological Applications 22: 1997–2006, http://smithsonianscience.org/?s=jay.

# APPENDIX 2. Status of Birds in Trouble

The status for each of the birds in trouble described in this book is given here using the terminology employed by various conservation organizations, as indicated in the notes for this appendix.

| Species | Status as designated by organization | | |
| --- | --- | --- | --- |
| | **IUCN** | **Audubon** | **ABC** |
| Trumpeter Swan | Least concern | Yellow | Vulnerable |
| Gunnison Sage-Grouse | Endangered | Red | At-risk |
| Greater Sage-Grouse | Near threatened | Yellow | Vulnerable |
| Greater Prairie-Chicken | Vulnerable | Red | Vulnerable |
| Lesser Prairie-Chicken | Vulnerable | Red | At-risk |
| Yellow-billed Loon | Near threatened | Yellow | Vulnerable |
| Clark's Grebe | Least concern | Yellow | Vulnerable |
| Black-capped Petrel | Endangered | Red | At-risk |
| Bermuda Petrel | Endangered | Red | |
| Ashy Storm-Petrel | Endangered | Red | At-risk |
| Reddish Egret | Near threatened | Red | At-risk |
| Ferruginous Hawk | Least concern | | Potential concern |
| Yellow Rail | Least concern | Red | At-risk |
| Black Rail | Near threatened | Red | At-risk |
| Whooping Crane | Endangered | Red | At-risk |
| Snowy Plover | Least concern | Yellow | At-risk |
| Piping Plover | Near threatened | Red | At-risk |
| Mountain Plover | Near threatened | Red | At-risk |
| Wandering Tattler | Least concern | Yellow | Vulnerable |
| Bristle-thighed Curlew | Vulnerable | Yellow | At-risk |
| Long-billed Curlew | Least concern | Yellow | Vulnerable |
| Hudsonian Godwit | Least concern | Yellow | Vulnerable |

| | | | |
|---|---|---|---|
| Red Knot | Least concern | Yellow | Vulnerable |
| Buff-breasted Sandpiper | Near threatened | Red | At-risk |
| Kittlitz's Murrelet | Critically endangered | Red | At-risk |
| Xantus's Murrelet (Scripps's and Guadalupe Murrelets) | Vulnerable | Red | At-risk |
| Craveri's Murrelet | Vulnerable | Red | At-risk |
| Ivory Gull | Near threatened | Red | At-risk |
| Flammulated Owl | Least concern | Yellow | Vulnerable |
| Spotted Owl | Near threatened | Red | At-risk to vulnerable (Mexican population) |
| Red-cockaded Woodpecker | Near Threatened | Red | At-risk |
| Red-crowned Parrot | Endangered | Red | At-risk |
| Black-capped Vireo | Vulnerable | Red | At-risk |
| Florida Scrub-Jay | Vulnerable | Red | At-risk |
| Island Scrub-Jay | Vulnerable | Yellow | At-risk |
| California Gnatcatcher | Least concern | Yellow | Vulnerable |
| Bicknell's Thrush | Vulnerable | Red | At-risk |
| McKay's Bunting | Near threatened | Yellow | Vulnerable |
| Colima Warbler | Least concern | Yellow | Vulnerable |
| Kirtland's Warbler | Near threatened | Red | At-risk |
| Cerulean Warbler | Vulnerable | Yellow | Vulnerable |
| Golden-cheeked Warbler | Endangered | Red | At-risk |
| Brown-capped Rosy-Finch | Least concern | Yellow | At-risk |
| Black Rosy-Finch | Least concern | Yellow | Vulnerable |

*Sources:* IUCN: Red List of Threatened Species (2011), ver. 3.1, http://www
.iucnredlist.org/; Audubon: Audubon 2007 Watchlist, www.audubon.org; ABC:
American Bird Conservancy, *List of the Birds of the United States with Conservation
Rankings (2012), www.abcbird.org.*

**STATUS DEFINITIONS**

IUCN categories: to be placed into a particular category, the species must meet at least one of certain carefully delineated criteria related to population size reduction, geographic range, population size, or probability of extinction. Critically endangered. Best available evidence indicates that the species meets at least one of five carefully delineated criteria and is therefore considered to be at extremely high risk of extinction in the wild. The criteria are: 80 to 90 percent population reduction over specified time periods, geographic range considerations (fluctuation, reduction, fragmentation), population size less than 250 and particular rate of decline, population of less than 50 mature individuals, and analysis indicating that the probability of extinction is at least 50 percent over ten years or three generations.

Endangered. Best available evidence indicates that the species meets at least one of five carefully delineated criteria and is therefore considered to be facing a very high risk of extinction in the wild. The criteria are: 50 to 79 percent population reduction, fewer problems in geography than for Critically endangered species, declining population size of less than 2,500, population size of less than 250 mature individuals, and analysis indicating that the probability of extinction is at least 20 percent over twenty years or five generations.

Vulnerable. Best available evidence indicates that the species meets at least one of five carefully delineated criteria and is therefore considered to be facing a high risk of extinction in the wild. The criteria are: 30 to 50 percent population reduction, fewer geographic concerns than the two endangered categories, declining population size of less than ten thousand, population size of less than one thousand mature individuals, and analysis indicating that the probability of extinction is at least 10 percent within one hundred years.

Near threatened. The species has been evaluated against the criteria but does not qualify for Critically endangered, Endangered, or Vulnerable now but is close to qualifying for or is likely to qualify for a threatened category in the near future.

Least concern. The species has been evaluated against the criteria but does not qualify for Critically endangered, Endangered, or Vulnerable.

Audubon Watchlist categories (a joint project between American Bird Conservancy and the National Audubon Society) :

Red. Species in this category are declining rapidly and/or have very small populations or limited ranges and face major conservation threats; these species are typically of global conservation concern.

Yellow. This category includes species that are either declining or rare; these species are typically of national conservation concern.

ABC (American Bird Conservancy) categories (2012) reflect a combination of scores for population size, range size, threats, and population trends.

At-risk. Indicates birds that need more urgent conservation attention; this category includes those most in danger of extinction.

Vulnerable. Indicates birds that deserve conservation attention; they typically have limited ranges, smaller populations, higher threats, or significantly higher declines (or a combination of these).

Potential concern. Indicates birds that may be regarded as currently safe but typically have smaller populations or ranges or slightly higher threats or more negative population trends than birds in the Secure category; local conservation actions may be appropriate, especially for those that congregate in large numbers at key sites.

Secure. Indicates birds that as a whole appear to have no immediate major conservation issues (if they are in decline and the decline were to continue, they may be upgraded to Potential concern).

| Species | Federal Status |
| --- | --- |
| Gunnison Sage-Grouse | Threatened |
| Greater Sage-Grouse | Candidate |
| Greater Prairie-Chicken (Attwater's) | Endangered |
| Lesser Prairie-Chicken | Threatened |
| Bermuda Petrel | Endangered |
| Whooping Crane | Endangered |
| Snowy Plover (western) | Threatened |
| Piping Plover | Threatened or Endangered, depending on population |
| Red Knot (rufa population) | Threatened |
| Xantus's Murrelet | Candidate |
| Spotted Owl (Mexican and Northern populations) | Threatened |
| Red-cockaded Woodpecker | Endangered |
| Red-crowned Parrot | Candidate |
| Black-capped Vireo | Endangered |
| Florida Scrub-Jay | Threatened |
| California Gnatcatcher (Coastal) | Threatened |
| Kirtland's Warbler | Endangered |
| Golden-cheeked Warbler | Endangered |

Birds not listed in this table do not appear on the published list in Title 50, chapter I, subchapter B, part 17, subpart B of the Government Printing Office's Endangered and Threatened Wildlife and Plants 50 eCFR 17.11 and 17.12, which is updated daily. The federal status shown here is as of January 16, 2015. The US Fish and Wildlife Service designates birds as Candidates for protection under the Endangered Species Act. See http://ecos.fws.gov/tess_public/pub/SpeciesReport.do?listingType=C&mapstatus=1; see also http://www.fws.gov/endangered/species/us-species.html.

# SUGGESTED READING

Askins, Robert A. *Restoring North America's Birds: Lessons from Landscape Ecology*. 2nd ed. New Haven: Yale University Press, 2002.

Barber, Lynn E. *Extreme Birder: One Woman's Big Year*. College Station: Texas A&M University Press, 2011.

Collar, N. J., and P. Andrew. *Birds to Watch: The ICBP World Checklist of Threatened Birds*. ICBP Technical Publication No. 8. Washington, D.C.: Smithsonian Institution Press, 1988.

Collar, N. J., L. P. Gonzaga, N. Krabbe, A. Madroño Nieto, L. G. Maranjo, T. A. Parker III, and D. C. Wege. *Threatened Birds of the Americas: The ICBP/IUCN Red Data Book*, part 2. 3rd ed. Washington, D.C.: International Council for Bird Preservation in cooperation with Smithsonian Institution Press, 1992.

Dunn, Jon L., and Jonathan Alderfer, eds. *National Geographic Field Guide to the Birds of North America*. 5th ed. Washington, D.C.: National Geographic Society, 2006.

Erickson, Laura. *101 Ways to Help Birds*. Mechanicsburg, Pa.: Stackpole Books, 2006.

Faaborg, John. *Saving Migrant Birds: Developing Strategies for the Future*. Austin: University of Texas Press, 2002.

Griggs, Jack L. *American Bird Conservancy's Field Guide: All the Birds of North America*. New York: HarperCollins, 1997.

Lebbin, Daniel J., Michael J. Parr, and George W. Fenwick. *The American Bird Conservancy Guide to Bird Conservation*. Chicago: University of Chicago Press, 2010.

Nigge, Klaus. *Whooping Crane: Images from the Wild*. College Station: Texas A&M University Press, 2010.

Poole, A., ed. *The Birds of North America Online*. Ithaca: Cornell Laboratory of Ornithology. http://bna.birds.cornell.edu/BNA/.

Rich, T. D., C. J. Beardmore, H. Berlanga, P. J. Blancher, M. S. W. Bradstreet, G. G. Butcher, D. W. Demarest, E. H. Dunn, W. C. Hunter, E. E. Iñigo-Elias, J. A. Kennedy, A. M. Martell, A. O. Panjabi, D. N. Pashley, K. V. Rosenberg, C. M. Rustay, J. S. Wendt, and T. C. Will. *Partners in Flight*

*North American Landbird Conservation Plan.* Ithaca: Cornell Laboratory of Ornithology, 2004. http://www.partnersinflight.org/cont_plan.

Rich, T. D., Coro Arizmendi, Dean W. Demarest, and Craig Thompson, eds. *Tundra to Tropics: Connecting Birds, Habitats and People; Proceedings of the Fourth International Partners in Flight Conference, February 13–16, 2008, McAllen, Tex.* Partners in Flight, 2009.

Sibley, David Allen. *National Audubon Society: The Sibley Guide to Birds.* New York: Knopf, 2000.

Stutchbury, Bridget. *The Bird Detective.* New York: HarperCollins, 2010.

Stutchbury, Bridget, and John Flicker. *Silence of the Songbirds.* New York: HarperCollins, 2010.

Terborgh, John. *Requiem for Nature.* Washington, D.C.: Island Press, 2004.

———. *Where Have All the Birds Gone? Essays on the Biology and Conservation of Birds That Migrate to the American Tropics.* Princeton: Princeton University Press, 1989.

Terborgh, John, and James A. Estes, eds. *Trophic Cascades: Predators, Prey, and the Changing Dynamics of Nature.* Washington, D.C.: Island Press, 2010.

Wells, Jeffrey V. *Birder's Conservation Handbook: 100 North American Birds at Risk.* Princeton: Princeton University Press, 2007.

# INDEX